Erratum, page 95.

Volvox, a few lines from the bottom of the page, should read *Vorticella*.

Artificers of Fraud

The Origin of Life and Scientific Deception

Peter Jones

Here's another one fit for burning! With best wishes,
Peter Jones.
15/5/13.

Orgonomy UK
PO Box 1331, Preston, PR2 OSZ.

Published by Orgonomy UK, PO Box 1331, PRESTON, PR2 OSZ,
March 2013.

ISBN 978-0-9574850-06

I dedicate this small book to all those scattered about our planet who are studying the life energy and furthering its natural functions. May it help to advance your labours.

For many have been frightened from the use of it, by imagining it required great skill in optics, and abundance of other learning, to comprehend it to any purpose: whereas nothing is really needful but good glasses, good eyes, a little practice, and a common understanding, to distinguish what is seen; and a love of truth, to give a faithful account thereof.

The likeliest method of discovering the truth, is, by the experiments of many upon the same subject; and the most probable way of engaging people in such experiments, is by rendering them easy, intelligible, and pleasant.

...but give me leave to request the favour of your concurrence, in my desire of explaining, to those that are ignorant, a science that may prove of the most eminent service towards the advancement of true knowledge, and in which everybody that has inclination and opportunity may be able to do something.

From the dedication to the president and fellows of the Royal Society of *The Microscope Made Easy* by Henry Baker, published in London in 1742. (Capital letters corrected in accordance with modern practice, punctuation and spelling as in the original text, Special Collections, Harris Library, Preston, Lancashire.)

Contents

	Introduction and acknowledgements	1
1	Robert Brown (1773-1858)	11
2	Wilhelm Reich (1897-1957)	24
3	Dismissing Reich	33
4	Rebellion from Below - The Background to Brown's Retraction	40
5	Motivations? The Assumptions of Mechanistic Science. Implications and Consequences.	51
6	The Drip, Drip, Drip of Deception	59
7	Repeating Brown's and Reich's Experiments	72
	Appendix 1: Brown's *Active Molecules* paper of 1828	113
	Appendix 2; Extracts from *The Bion Experiments on the Origin of Life*, *The Cancer Biopathy*, and *Ether, God and Devil* by Wilhelm Reich	132
	Books by Wilhelm Reich	137
	Suppliers and Equipment	138
	Orgonomy Today and Further Study	143
	Glossary	145
	Index	153

Illustrations between pages 80 and 81

C O R E's Olympus BX50 microscope and camera
Brown's microscope
Brunel SP 100 microscope and improvised work area
Brunel SP 150 microscope
Wilhelm Reich (1897-1957)
Robert Brown (1773-1858)
Frontispiece, Needham's *Nouvelles Observations Microscopiques* (1750)
Needham's illustration of pollen grain erupting under water
Sample page from Brown's slips
Bions from rock-dust
Bionous forms in hay infusion
Bionous, granular matter from bursting freesia pollen in water
Grass infusion experiment – slide
Bions emerging on grass surface
Bionous swelling on edge of grass blade

Introduction

In this book I unearth and expose facts first revealed in 1828 by the great botanist and microscopist, Robert Brown, and sequestered by science ever since. The most important of Brown's pioneering observations have been ignored and concealed and their reality forgotten since he reported his findings in 1828.

To complete our picture, we leap from 1828 to the nineteen-thirties, when Wilhelm Reich was doing his pioneering research that led to his discovery of the bions and the process of bionous disintegration. Reich touched on Brown's work and went far beyond it, though this in no way belittles the greatness of Brown's achievement. Reich's findings, too, have been ignored to this day and are still hidden from the world at large.

'And who is Reich?' most readers will ask. And what are bions and bionous disintegration? What have they got to do with Robert Brown? Who's he, anyway? If you read on, I will tell you. You could, of course, read Reich's own books on the subject. These are at present available to buy, and can anyway, be borrowed from public libraries, if you do not think the question is worth spending forty or fifty pounds on. Brown's works are much harder to come by and are buried in small anthologies in library stacks. They are still accessible to the determined ferret. The relevant text, Brown's essay on *Active Molecules*, is reprinted in this book. (See page 113.) As far as I know, there is no printed version available at present. It was recently (2004) re-published in a six-volume anthology by Thoemmes Press.[1] This anthology is now out of print.

I am publishing this book at my own expense for simple reasons. Firstly, no mainstream publisher in the UK will touch a positive book about Wilhelm Reich at the moment, when his reputation is as low as it has ever been in this country since his death in 1957. Secondly, it is important to present this information to the world while I am still able to. Thirdly, a desperate hope this, in the hope that the profits from sales will accrue to C O R E and help provide the building that is so urgently needed to house our priceless library and scientific equipment, not to mention a home for orgonomic teaching and research in this country. Thank you for your contribution, if you have bought this book.

So… what is this book about?

From 1934 to 1939 Wilhelm Reich worked in exile from Nazi Germany in Oslo, Norway. He was conducting experiments that showed him a process till then unnoticed by science, *bionous disintegration*. This occurred when dead vegetable matter swelled in water and started to break down into tiny, highly charged vesicles which had some of the characteristics of the living. Agglomerations of these would sometimes gather together to form primitive proto-organisms. He went on to discover that the same process of swelling and disintegration into these highly charged vesicles occurred with ground, inanimate materials such as sand, ash, coal, or iron filings. He named these vesicles *bions*.

Any suggestion that 'life', however primitive and simple, or even forms transitional between non-life and life, can originate today from inanimate matter is laughed out of court by modern biology and not investigated scientifically. It is dismissed, in the same way that politicians dismiss ideas from other parties, as unthinkable. The label of dismissal that was thrown at Reich and which is still thrown at his work and anyone who tries to repeat his experiments is 'Brownian Motion'. Some readers may dimly remember the phrase *Brownian motion* from their compulsory science lessons at school. This is the territory of conventional science, after all. Brownian motion, as now understood, is the motion imparted to tiny particles of material in a fluid by the kinetic energy of the moving molecules of this fluid. This random motion only affects particles below a certain size. Once a particle is above a certain size, about 1µm in diameter (one millionth of a metre), the motion disappears. Brownian motion can even be seen in smoke particles suspended in air.

So this is what Brown was describing in his article on *Active Molecules*? Now…that's the big question. This is what all biologists and physicists think Brown was describing in his paper. This is taken for granted in all textbooks. But, if you have studied Reich's writings, even better, if you have repeated some of his basic bion experiments, and take the trouble to read Brown's actual article, not the neutered versions of it that appear in textbooks, you will realise that Brown's words are immediately familiar. Can it be? Yes, it can. He is! Brown is describing the bions. Without realising it, he has hit upon the bions. With no bio-energetic framework to place his findings in, he is, of course, quite lost. However, he is a real scientist, so what does he do? He simply records his findings and leaves it to others, maybe posterity, to work out what he has been looking at. He describes pulsatory and contractile motion in some of the particles he is observing. These comments are ignored. Other scientists observe what they assume he was describing and eventually along comes posterity in the form of Albert Einstein, who does some difficult sums, and explains the motion apparently observed by Brown in a way that hardly anyone else can follow, because they are not advanced mathematicians. His findings are, apparently, confirmed by practical observation by a French mathematician, Jean Perrin.[2] And that is that. Brownian motion has been explained. So Reich was wrong, after all, and we can all go home and sleep peacefully. (Phew! That was a narrow squeak!) After all, there is no life-force beavering away before our noses and life is definitely not springing into being under our very feet in wet soil or little puddles.

There are some snags to the widely accepted explanation of Brownian motion that we see in most textbooks. Brown actually refers to contractile movements in the particles he is observing.

> …their motion consisting not only of a change of place in the fluid, manifested by alterations in their relative positions, but also not infrequently of a change of form of the particle itself; a contraction or curvature taking place repeatedly about the middle of one side, accompanied by a corresponding swelling or convexity on the opposite side of the particle. In a few instances the particle was seen to turn on its longer axis.

So that is most definitely not Brownian motion, as described in the textbooks, is it? (I shall come back to this apparent contradiction in more detail later.) The other snag, a snag from the point of view of this book, is that soon after writing his first report, published in 1828, Brown proves himself less of a scientist than I had at first thought: he withdraws his assertion that his 'active molecules' are possibly alive and a basic building element of life. In the original article he compares his findings to those of earlier biologists, in particular De Buffon and Needham.

Here is what he writes about the possible connection between his own observations and the earlier work of other biologists:

> Reflecting on all the facts with which I had now become acquainted, I was disposed to believe that the minute spherical particles or Molecules of apparently uniform size, first seen in the advanced state of the pollen of Onagrariae, and most other Phaenomagous plants, - then in the antherae of Mosses and on the surface of the bodies regarded as the stamina of Equisetum, - and

> lastly in the bruised portions of other parts of the same plants, were in reality the supposed constituent or elementary Molecules of organic bodies, first so considered by Buffon and Needham, then by Wrisberg with greater precision, soon after and still more particularly by Müller, and, very recently, by Dr Milne Edwards, who has revived the doctrine and supported it with much interesting detail. I now therefore expected to find these molecules in all organic bodies; and accordingly on examining the various animal and vegetable tissues, whether living or dead, they were always found to exist; and merely by bruising these substances in water, I never failed to ascertain their apparent identity in size, form, and motion, with the smaller particles of the grains of pollen.

Here are the main sentences of his *Additional Remarks*:

> In the first place, I have to notice an erroneous assertion of more than one writer, namely that I have stated the active molecules to be animated. This mistake has probably arisen from my having communicated the facts in the same order in which they occurred, accompanied by the views which presented themselves in the different stages of the investigation; and in one case, from my having adopted the language in referring to the opinion of another enquirer into the first branch of the subject.

The tone of these 'additional remarks' is strange. He appears to have changed his own mind about his discovery. There is a clear contradiction in the statements of the two sections. To me, it is obvious that someone has warned him off his first claim, possibly advising him that his career would be in jeopardy, if he were to publicly cling to such a point of view. Readers can judge for themselves. The full text is in appendix 1. It seems relevant to this apparent change of mind that Brown had no income of his own and was at the mercy of the good graces of the high and the mighty of science at that time. He very much needed to keep his post at the British Museum and not to fall out with the scientific establishment of the day, though this assumption is only my own informed guess. There are also strong political and psychological reasons why Brown may have withdrawn his original opinion. See chapters 1 and 4.

The tone of these contradictory remarks reminds me of a contemporary politician caught out by a persevering interviewer on a news programme. 'Well…that is not what I really meant. You are taking my remarks out of context.' It is a smoke-screen, constructed, while he squirms his way out of the hole that he has landed in.

So far, I have been talking of what Brown wrote. What do we find if we repeat his experimental observations? No scientists will want to argue with experimental findings, will they? If we repeat his investigations we find that much of the motion that he had seen was definitely not Brownian motion. It was *bionous motility*. (We can observe Brownian motion and bionous motility in the same preparation.) We shall come to that in more detail later. If you can't wait, please see chapters 1 and 8.

So…for more about Robert Brown and his work, please see chapter 1: for more about Wilhelm Reich, see chapter 2.

Repeating the experiments on which these arguments are based may seem to be a long way beyond the possible to armchair readers who have no experience of practical science or microscopy. In fact this is not as difficult or expensive as you may imagine. For more details see chapter 7. Please do not rule this out. This work is very interesting indeed and a project full of life that will readily draw in anyone whose own life-energy has a bit of sparkle left. These are not investigations that you have to make yourself do. They will really get you out of bed in the morning. If you have access to an 'A' level biology lab, you will be able to repeat these experiments. If you do not have this access but can collect a few hundred pounds towards the purchase of a microscope and the few extras needed for these experiments or already have such a sum and are willing to spend it on this project, you can get on with it on your own. There is enough basic information in this little book for you to repeat these experiments. A few hundred pounds sounds a lot, but people often spend such an amount on a camera or a computer. A microscope needs very little maintenance and lasts a lifetime, even several lifetimes, if you look after it carefully. You can buy a good quality second-hand microscope on ebay for only £200-300 which will be quite adequate for all this microscopical work. I have in the last few weeks checked many of my findings on exactly such an instrument, a Leitz SM Lux, which cost £300. If you come to one of C O R E's bion workshops you may well end up using the same microscope yourself. If you don't recognise the name Leitz, I must tell you that they were one of the top-grade microscope makers in the world, possibly the best, and now trade as Leica. You can buy a new microscope which will enable you to repeat these experiments for less than £500. (Since I wrote this paragraph, I have bought a new microscope on ebay for £145. Obviously it is a Chinese-made model at the low end of the market, but its optics are excellent and it has proved more than adequate for these experiments.) It has now been sold on to one of C O R E's few active supporters and the new owner is himself repeating the bion experiments.

The Right to Criticism

This book, if it attracts attention, will be heavily criticised. To have any right to criticise it, critics must have read Reich's research reports, have repeated his experiments, have read Brown's 1828 paper, and repeated his pollen investigations. Doubtless I will be criticised by people who have done none of these. Some of Reich's critics refused to look down his microscope for fear of what they might see. I expect the same refusal now.

Personal Background

What gives me the right to tell this story? By the lights of conventional science – nothing. I have no scientific qualifications whatsoever and have passed only one science examination in my life. But I have been a serious student of orgonomy, the science of the life energy, for over 40 years now. I bought a microscope in 1997 with the intention of repeating Reich's bion experiments, something that I had wanted to do since I first read about them in the nineteen sixties. This instrument was a high-quality research grade microscope and has proved a wonderful aid in this work, easy to use, accurate in its optics and engineering, and very forgiving to an amateur with limited experience. It is a bizarre thing for an amateur to have and use such a high-quality instrument. There can't be many such microscopes in the country without the facilities of a professional laboratory around them. My choice of this expensive Olympus model was governed by a mixture of ignorance and determination about my intentions. I have never regretted it, not for one minute. You do not need a degree in biology or microbiology to get good results with a microscope. I include quite a lot of information on how to use a microscope for these experiments, but almost anyone with moderate levels of vision, intelligence, and manual dexterity, if presented with a microscope, would soon be able to work out how to use it just by looking at it.

Over the years since this purchase I have repeated Reich's basic bion experiments and eventually Brown's experiments, too. More recently I have bought, on behalf of C O R E for teaching purposes, much cheaper microscopes, partly so that we had extra instruments for our bion workshops and conferences, but also to find out just what the cheapest microscope is on which one can repeat these experiments. The microscopes in university research labs or hospital labs cost several thousand pounds and one can spend twenty thousand pounds and more on such instruments, if one goes in for expensive contrast systems such as Nomarski DIC (differential interference contrast). So we are threatening Goliath with a pea-shooter, using microscopes at the bottom end of the market. But the optics on imported Chinese models are excellent for the price and take a student through these experiments. I have repeated them on a Brunel trinocular SP100, (price about £580 at the time of writing). You can buy equally good second-hand microscopes on ebay for half this sum.

Whenever I have been writing about these experiments I have repeated what I am describing to be sure that I have forgotten no details and not let my imagination or memory run away with me. Visiting students have repeated my investigations and confirmed my findings. Reich's experiments have been repeated by other orgonomists, long before I started on them, and written up and developed in the orgonomic literature. I have not come across or heard of anyone who has repeated Brown's experiments and investigations recently apart from Dr Brian J Ford. (See chapter 6.) Most writers referring to Brown's findings and Brownian motion simply repeat the neutered version handed down in the history books. This version says that he was the first observer to notice and describe what is now known as Brownian motion while observing pollen grains in water.

It is a great educational experience to repeat the experiments of the pioneers of microscopy from their own writings. If I were a professional teacher of microscopy, I would suggest students replicate these experiments as part of a course. If we ever run courses in orgonomic biology, I would certainly insist that students do this. C O R E's recent booklet, *Further Experiments with the Microscope for the Amateur Orgonomist*, contains information on experiments by van Leeuwenhoek (1632-1723), Spallanzani (1729-1799), Brown's pollen investigations of 1828, and an investigation by H C Bastian (1837-1915.

Acknowledgements

Most books of popular science or serious non-fiction usually include pages of thanks to organisations who have provided funds to support the author during their labours and to droves of colleagues who have helped the writer in their work. These thanks often run to several pages. I wish it were the same in the field of orgonomy. Alas, the path of the orgonomic scholar and researcher in any country, but especially in the UK, is a lonely and stony one. The writing of this book has not been supported financially by anyone. It has been a labour done for the sake of its importance and has been a huge drain on my pocket. If anyone has supported it, it has been the British taxpayer, who paid my modest NHS salary until I retired three years ago. Ten years of this support full-time and five years part-time have allowed me to finance serious practical orgonomic research with relative ease. So, thanks to the UK taxpayer.

Thanks, too, to the small band of orgonomic colleagues, mostly in the USA, who have given generous moral support to this project and suggested directions and ideas and connections that I might not have seen without their aid. In alphabetical order they are: Philip Bennett, Dean Davidson, James DeMeo, Bernd Senf, Maxwell Snyder, Stergios Tsiormpatsis. Thanks to the Wilhelm Reich Infant Trust Fund and Reich's executor Mary Higgins and the late Roger Straus of Farrar, Straus and Giroux, without whose indefatigable efforts the works of Wilhelm Reich would not still be in print. If his writings were not now easily available, my work would have been enormously more difficult and laborious.

Thanks to Judith Hooper, author of *Of Moths and Men*.[3] Her exposé of the fraudulent research that appeared to justify the Darwinist interpretation of the rise of the melanic form of the peppered moth inspired me to undertake this project. The theft of Brown's original findings seemed to me even more disgraceful than Bernard Kettlewell's outrageous bending of his dubious evidence.

A special thanks to Dr James Strick, lecturer in the history of science at Franklin and Marshall College, Lancaster, Pennsylvania, for information on the historical background to the reception of Brown's discovery. When he realised that I was working on Brown's experiments and their significance for Reich's bion research, he very kindly sent me a copy of his unpublished essay, *Robert Brown and Brownian Movement: Radicalism, Spontaneous Generation and Microscopy in Nineteenth Century England*. It was wonderful to receive unsolicited a paper so relevant to this work and so full of fertile leads. Chapter 4 rests very much on his paper and would have been greatly inferior without his input. His contribution, that of a professional historian of science, has been enormous.

Thanks to Margaret Troy, who told me about hive-pollen for sale and to Ann Benson for finding me some evening primrose flowers, so that I could repeat Brown's investigation accurately. Thanks to my brother Jeremy for help with my endless computer problems, for helpful design suggestions, and for patiently listening to my grumbles about the vagaries of the digital world. Thanks to my sister Pamela, whose emotional support has never crumbled.

Thanks to Michelle Sturman and Iain Stobbs who have generously read and corrected the proofs of this text. They have done an outstanding job and eliminated many errors of mine. Thanks to Carol Henshaw for finding Michelle as a proof-reader. Needless to say, I take responsibility for any remaining errors and a few imperfections that I decided to preserve according to my own editorial judgement.

Thanks to Dan Clement for his cover design. Thanks to the Natural History Museum, London, and the Linnean Society of London for permission to use photographs. Thanks to Farrar, Straus and Giroux, New York, for permission (at a price) to reprint extracts from Reich's works.

A final, and perhaps the biggest thanks of all, must go to the staff at so many reference libraries. An independent researcher with no academic support depends completely on these workers for access to reference works and rare documents. It is a miracle in our commercialised world that these reference libraries and collections still exist and provide access to valuable and rare items to anyone who needs it. These institutions are truly democratic, open to anyone who wishes to use them. You do not have to have a PhD in biology, not even a BSc or BA, to be welcomed into their world. Throughout the writing of this book and the long period I spent digging over the ground before the idea finally took shape I spent much time in the following libraries and am very grateful to their staff for support provided; Lancaster City Library, Hull Central Library, the Manchester Central Reference Library; the reference library of the Harris Library, Preston, where librarian Claire Sutton has been particularly helpful, as have other librarians there; the inter-library loan system via the Harris Library, Preston; the rare manuscripts section of the library of University College, London; and last but not least, the library of the Natural History Museum, London, whose librarians, Armando Mendez and Natalie Pope, gave me very helpful and willing assistance in my quest to try to track down exactly what Robert Brown saw and wrote while he was discovering what has become known as Brownian motion. Except that, as you will see, it was not what we now call Brownian motion at all, but bionous motility that he had discovered. I am still a frequent visitor to the Harris Reference Library in Preston and am every week, sometimes every day, grateful to the librarians there for their tireless work on my behalf. It is a matter of great concern to me and distress to the staff there that the future of this great library is threatened by plans for re-organisation of the Lancashire library service.

The staff of these libraries are really interested in their work. They scratch their heads to help you solve a difficult problem, and often come up with the most inspired answers. There is none of that 'if it's not on the shelves we haven't got it' attitude. If it is not on the shelves, they give serious thought to getting hold of it or to finding a way of obtaining the information in some other form. These libraries are a joy to use.

PS 2010: Preston's magnificent, thriving reference library is no more. Large numbers of basic reference books have vanished into the basement and much of the research for this book carried out in that library would no longer be possible today.

Abbreviations of the titles of Reich's main works used in the references. See page 137 for bibliographical details of these works. The first date is that of original publication, the second the date of publication of the currently available English-language edition.

CA - Character Analysis (1945, 1972)
BISA - The Bioelectrical Investigation of Sexuality and Anxiety (1938, 1985)
BEOL - The Bion Experiments on the Origin of Life (1939, 1979)
FO - The Function of the Orgasm (1942, 1983)
CB - The Cancer Biopathy (1948, 1974)
EGD - Ether, God and Devil (1949, 1973)
CS - Cosmic Superimposition (1951, 1973)
SW - Selected Writings (1960, 2002)
GTTN - Genitality in the Theory and Therapy of Neuroses (1927, 1985)
PY - Passion of Youth (1988)
BP - Beyond Psychology (1994)
PIT - People in Trouble (1953, 1976)

[1] Strick J (ed) (2004); The Origin of Life Debate: Molecules, Cells and Generation, Thoemmes Continuum, Bristol.

[2] Perrin J (1923); Atoms, Chapters III and IV, Constable, London.

[3] Hooper J (2002); Of Moths and Men, Fourth Estate, London.

Chapter 1 Robert Brown (1773-1858)

> In many of the substances examined, especially those of a fibrous structure, as asbestos, actinolite, tremolite, zeolite, and even steatite, along with the spherical molecules, other corpuscles were found, like short fibres somewhat moniliform, whose transverse diameter appeared not to exceed that of the molecule, of which they seemed to be primary combinations. These fibrils, when of such length as to be probably composed of not more than four or five molecules, and still more evidently when formed of two or three only, were generally in motion, at least as vivid as that of the simple molecule itself; and which from the fibril often changing its position in the fluid, and from its occasional bending, might be said to be somewhat vermicular. (Robert Brown, *Active Molecules*, 1828.)
>
> The concept of Brownian motion arose from his observation that very fine pollen grains suspended in water move about in a continuously agitated manner. He was able to establish that inorganic materials such as carbon and various metals are equally subject to it, but he could not find the cause of the movement (now explained by kinetic theory). (Extract from Hutchinson Dictionary of Scientists, Helicon Publishing, Oxford, 1996.)

Although known to the scientifically educated as the discoverer of Brownian motion and the nucleus of plant cells, Brown is hardly known outside science. He does not have the charisma or reputation of Galileo or Darwin. From all accounts he was a shy, scholarly man, totally devoted to his scientific pursuits, not the sort of scientist who could easily be made into a public hero or a controversial figure. (But that's not the whole story, because a repellent persona has not stopped Isaac Newton becoming a hero of science with statues in museums.) There is still, as far as I know, only one full biography of Brown.[1] This is a very botanical and technical biography, concentrating on the species he 'collected' and classified and the papers he wrote. It includes detailed comparisons with predecessors and the scientific assessments of other workers. It contains very little personal material about him and he remains a shadowy figure even after a reading of that biography. It may be, of course, that there is little to know about him.

He seems to have been shy, diffident, and nervous about his own achievements to the point of complete self-concealment. For the purposes of this book we do not need to know about him as a person, though I feel that a greater knowledge of him as a man would make my interpretation of his *Active Molecules* article more certain and help us to understand better the reasons for his change of heart in his *Additional Remarks*.

Brown was born in Scotland in 1773. His father was one of the two last recalcitrant bishops of the Episcopalian Church of Scotland, subject to the severest restrictions on his preaching activities for his refusal to swear allegiance to George III on the death of the Stuart pretender, Prince Charles, in 1788.[2] He studied at Edinburgh University in the heyday of the Scottish Enlightenment, when the capital was blossoming intellectually and academically. He trained as a physician, but his first interest was always botany. He is famous, far more famous than for his discovery of Brownian motion, in the world of botany for his prodigious collecting work on a long voyage to Australia from 1801-1805 and his later discovery of the nucleus in plant cells.[3] He was one of the first European natural scientists to classify and catalogue the flora of the newly colonised continent, which was being explored and mapped at the same time by the expedition of which he was a member. He was following on with the work started by another great British botanical collector, Joseph Banks (1743-1820), who had earlier sailed with Captain Cook on the *Endeavour* (1768-1771) as the expedition's naturalist.[4] At this time, when the British navy was exploring and surveying the world, it was common practice for the expedition to take with it a naturalist. He would typically be an educated gentleman of independent means who could pay his own way on the journey, though Brown actually had a paid appointment. There were at the time no formally acknowledged and trained scientists and the word was not in use even by then. Charles Darwin (1809-1882) also sailed as the naturalist on another surveying expedition and collected much of the information for his *Origin of Species* during that expedition.[5]

Brown became Banks's private librarian for his last ten years from 1810 to 1820.[6] Banks had a large private fortune and Brown had to earn his way in the world. He found himself in a socially subservient position to the gentlemen scientists of his day. This may have influenced his ability to stand up for unpopular or controversial opinions and to be seen to take a stand against the prevailing spirit of the times. Brown is a shadowy figure and it is difficult to know what he thought and felt. He had to work for his living, unlike Banks and Darwin, and became botany keeper at the recently founded British Museum after working as Banks's private librarian.[7] (Banks had bequeathed his enormous collection to Brown for the duration of his life and then to the British Museum and Brown moved with it there after Banks's death.) Brown still found time to do a great deal of research and writing and left a large body of work at his death. He made important contributions to the classification of plants and made many discoveries concerning plant reproduction as well as the two important discoveries that he is famous for – Brownian motion and the nucleus of plant cells.[8]

A paradox of this story is that Brown, a fastidious, shy bachelor, was, barely realising it, working in the field of sexuality, just as Reich was. Reich's discovery of the orgone, the bions, and bionous disintegration all stemmed from his study of the orgasm function and his discovery of orgonotic pulsation. This was no coincidence. He said that no other field of study could have led him to these discoveries so effectively.[9]

I think Brown deserves to be much better known. (He is, needless to say, a great name in the world of plant sciences.) He has been unlucky in that botany, or plant sciences, as it is called now, has no glamorous place in the history books of science. Most people have heard of Gallileo, Darwin, or Newton, and have a vague idea of what they discovered. There are any number of biographies and potted versions of their life-work, ranging from Ladybird Books to massive academic texts, but the great pioneers of botany - Grew, Ray, Hales, and Brown, remain unknown except within the world of plant studies. If science had followed a different direction as it developed in the early nineteenth century, Brown might have been far, far more famous than he is now, as the first scientist to observe and describe bionous disintegration and the bions, the transitional forms between the living and the non-living.

As with many discoveries made far before their time, there was no framework at all into which Brown's discovery could be placed and so it floated in limbo for decades and eventually was assimilated by physics into conventional science as something that it was not. I was going to say that no-one in 1828, when his *Active Molecules* article was first published, could conceivably have realised that these items were bions and that there was a charge of life energy driving their movement. But that does an injustice to Brown. He, himself, realised this at first, writing that these were possibly the basic elements of life and made connections between the items that he had observed and those hypothesised or observed by earlier scientists.[10]

Readers new to the history of science will be surprised to know that it is possible for scientific discoveries to be made and then to be ignored. This does occur when a discovery is 'ahead of its time' or completely opposed to the popular currents flowing in science at any given time. Ludwik Fleck made this point as early as 1935.[11] Brown has a great predecessor in his own field, that of microscopy, one whose work was so advanced and so little understood, that other workers were unable to repeat his intricate investigations. Anton van Leeuwenhoek (1632-1723), the great Dutch pioneer of the microscope, was the first scientist to see, describe, and classify micro-organisms using a microscope.[12] He was a genius at making and using lenses and observing, and few other people were able to repeat his investigations in his lifetime.[13] His discoveries were more or less ignored for another hundred years. Even in his own lifetime critics questioned his observations because they were unable to repeat them themselves. A theory concerning the causes of tuberculosis that apparently corresponded completely with modern theories of the contagious nature of the disease was published in England as early as 1720 by a doctor Marten.[14] This, too, languished unnoticed and forgotten. Other workers, generations later, made the same connections. Microscopy, too, stood still until it was taken up again by other scientists in the later eighteenth century and early nineteenth. Leeuwenhoek himself made no connection between the tiny organisms that he had seen and infectious illnesses. No-one at the time made the connection either, and it was not until the mid-nineteenth century that the germ-theory of disease began to be even considered as a viable explanation for illnesses, even though workers over the centuries from the early Greeks onwards kept occasionally putting forward the theory that invisible little animals caused diseases.[15]

Reich's background in psychology, psycho-analysis, and bio-energetic psycho-therapy was needed before anyone could give Brown's discovery, (of the bions, bionous disintegration, and bionous motility), its rightful place in science and make the connections between it and surrounding knowledge. I sense the same lack of a context in the work of H C Bastian (1837-1915), who also did a great deal of work on the origins of life and who also must have seen bions in his preparations without realising what they were.[16] Needless to say, neither Brown nor Bastian made any connections between their discoveries and human emotions or sensations. Reich was the first to realise that there is a connection between animal (including human) emotion and sensation and these primitive life-forms.[17] Now that may be interesting to readers. What a paradox that Brown, who seems to have been such a compulsive and inhibited character, should stumble upon a manifestation of the life energy and recognise it as such to start with.

Lay readers on the fringes of science will be delighted to read that one scientist at least, Wilhelm Reich, has had the courage to claim that there is a connection between the motion of nature and emotion and sensation within ourselves. Modern science resolutely denies such a connection. The denial of this connection and lack of any awareness of it within themselves is an important part of the unconscious rationale of modern science and scientists.

I base this claim on the fact that any conventional scientist shows an apparently inexplicable fury whenever any scientific work gets even close to this world, the world of life energy and its functions within nature. He gets even more furious when the work becomes active in this area and comes up with serious experimentation, discoveries, and testable claims. It is impossible to have a serious debate on these matters because orthodox science refuses to look the matter in the face.[18]

But enough of commenting and opinion. Let us get down to the meat of this book. What did Brown actually say in his famous article? It is included in this book in Appendix 1, so you can see for yourself, if you wish. Please read it. In all these debates there is no substitute for the original writings of the workers involved. The quotation at the head of this chapter is as good a summary as one can find of the orthodox version of Brown's discovery. Our investigations will show a great chasm between this version and what Brown actually wrote and an even wider chasm between what he observed and the orthodox version of what he observed.

I shall summarise the important parts here on the assumption that readers will read the original article in its own place in the book.

Brown starts by describing the lens and microscope that he has used for his observations, so that other workers can repeat his observations. He describes his first observation that revealed to him the vivid motion of various types of pollen grains in water. 'Grains' here means the microscopic particles that emerge from a pollen sac (also commonly called a grain) when it bursts in water and releases granular matter. He tests this initial observation by repeating the test with many different types of pollen. He finds that pollen grains from many different species of flowers show the same movement.

To further check his observations he tests pollen from plants that have been dead for many years and pollen that has been preserved in spirit for years, too. In all cases, he finds the same movement:

> …specimens of several plants, some of which had been dried and preserved in an herbarium for upwards of twenty years, and others not less than a century, still exhibited the molecules or smaller spherical particles in considerable numbers, and in evident motion, along with a few of the larger particles, whose motions were much less manifest, and in some cases not observable.[19]

These extensive checks show just what a true scientist Brown was in that he tested his findings to the nth degree to make sure that they were not a random finding, not a fluke. The comment on a relationship between the motion observed and the size of the particles is one of the few that suggest that at times he was observing true Brownian motion. This is only seen in particles below a certain size. It also suggests that he was observing particles derived from the pollen grains rather than the pollen grains themselves. He repeatedly in his own handwritten notes ('slips') refers to whether the pollen has burst or not. Once a pollen sac bursts under water it throws out both pollen grains and a large amount of granular material which produces bions very quickly.

I have made it my main interest to repeat Brown's investigations as best I can and as exactly as I can, and to determine exactly what he did see while he was doing these experiments. I have gone as far as going to the botanical library at the Natural History Museum in London to read his own, original notes. Like his diaries, these do not give a lot away. As I read his slips, (those that are decipherable, that is, as his writing is difficuilt to decipher), I recognised whole sentences which are found almost unaltered in his final article. I came across an expression that figures in the article and which no-one else, as far as I know, has quoted. He describes the motion of some of the particles he is observing as 'slightly vermicular,' in plain English, worm-like. This must surely be a reference to the contractile or pulsatory movement that he has noticed in some of the particles, bulges forming on one side and, concavities on the other, and the undulating motion of the agglomerations of the molecules.

As readers will see if they turn to the original text, once Brown was sure that all species of pollen that he tested showed this motion or produced particles that showed the motion, he moved to test inanimate materials for the same movement. He found that the dust or soot in London also consisted of these motile particles. A sample of fossilised wood that would still burn produced large quantities of the molecules, as did even older, fully fossilised wood, after it had been 'bruised', (Brown's own word). It does not occur to Brown that these molecules might be the product of a process. He simply assumes that they are contained within the materials that he is testing.

His next step is to test completely inorganic materials to see whether they produce the molecules, too. He finds that all the mineral materials locally available and common metals, even ground glass, also produce the molecules.

> …In a word, in every mineral which I could reduce to a powder, sufficiently fine to be temporarily suspended in water, I found these molecules more or less copiously; and in some cases, more particularly in siliceous crystals, the whole body submitted to examination appeared to be composed of them.[20]

Siliceous crystals are silica compounds and Brown's findings concerning these substances incidentally confirm my own orgonomic findings that ground clays, feldspar, and mica, all of which contain large amounts of silicates, produce bions easily and in large quantities. Other incidental comments of Brown's confirm other orgonomic findings, for instance, Reich's that dead grass breaks down into bions more quickly than green grass picked in the spring.[21] Anything that negatively affects the aliveness of vegetable matter damages its capacity to resist the break-down process. Brown's bruising will have a similar traumatising effect on the grass tissue and accelerate its bionous disintegration.[22] (Brown does not say, as far as I could see from his notes, exactly what he did to bruise his samples. In my own repetition of his experiments I squeezed samples between the blades of a pair of surgical forceps and clamped them closed, as one can with such an instrument.)

I shall leave the details of how we repeat some of Brown's experiments and investigations to a chapter on that subject alone. A how-to-do-it section would interrupt the flow here. Please just take as read for the time being the information presented here. The practical chapter will go into details. What do we find, if we read this article carefully and repeat some of the investigations carried out by Brown himself? Let us try the one with two samples of grass, one with grass blades that have been bruised and one with grass blades that have simply been cut, both samples then being immersed in boiled water. The findings, after a few days, exactly confirm Brown's claims. The same goes for ground coal, glass, clay, and mica.

As Brown says, almost any material that can be ground up finely produces his molecules,[23] what I will call bions. In my improvised laboratory facilities in my home I have got a book-case in which I keep any material that I have tested for bionous disintegration. The top two shelves hold more than seventy different samples of commonly occurring materials that I have ground up and tested and a small number of less common ones that I have collected samples of on holiday or been given by friends who know of my interest. A few are rocks or minerals not commonly found in this country, which I have bought from geological suppliers

An interesting finding of my own, which I obtained without understanding it, also incidentally confirms the fact that Brown's molecules are frequently bions and not always particles showing Brownian motion. Before I started my bion experiments I bought a pH meter, as it struck me intuitively that it would be relevant and important to assess the pH levels of any bion cultures that I prepared. I discovered that, as I ploughed my way through every possible material that I could test, some items produced either no bions at all or ones with a very short life, sometimes only a few minutes or so. These were those from a small number of cultures with a high pH value, around 12.0, that is, very alkaline or basic preparations.

I then came across an article by two German orgonomists, Palm and Döring, drawing a general conclusion from similar observations that they had made while testing sea-sand from Sri Lankan shores for bionous growth. Like English limestone, the sand there is of animal origin, and produces preparations with a high pH, which do not produce bionous growth. They come to the conclusion that bionous growth does not occur above a pH level of about 12.[24] It occurred to me to test this idea by seeing if, for example, limestone would produce bions in a preparation that was less alkaline than the one we get if we use water. I made the same preparation, only using household vinegar rather than water. (This preparation had a pH of about 4.0.) It produced generous bion growth.

Brown makes a discovery also made by Reich later, that soot and coal produces his active molecules, what Reich would later call motile bions. He repeatedly writes that these materials *contain* the molecules.

> I examined also various products of organic bodies, particularly the gum-resins, and substances of vegetable origin,

extending my enquiry even to pit-coal; and in all these bodies Molecules were found in abundance. I remark here also, partly as a caution to those who may hereafter engage in the same inquiry, that the dust or soot deposited on all bodies in such quantity, especially in London, is entirely composed of these molecules.[25]

Brown concludes that almost everything that can be ground up produces his active molecules. It would be tedious and repetitive to quote his findings word for word concerning every sample that he tests. Below are some short comments concerning the many materials that he has tested.

...a specimen of fossil wood...in a state to burn with flames...I found these molecules abundantly...a minute portion of silicified wood...molecules in all respects like those...were readily obtained from it...the whole substance of the petrifaction seemed to be formed of them...I inferred that these molecules were not limited to organic bodies, nor even to their products...to establish the correctness of the inference and to ascertain to what extent the molecules existed in mineral bodies, became the next object of enquiry...a minute fragment of window glass...I readily and copiously obtained molecules agreeing in size, form, and motion...several of the simple earths and metals...rocks of all ages...yielded the molecules in abundance...Their existence was ascertained in each of the constituent minerals of granite...In a word, in every mineral which I could reduce to a powder...I found these molecules more or less copiously...[26]

If we repeat Brown's experiments with, say, ground glass, various metals, granite, basalt, volcanic rock from the Canaries, and so on, the results are familiar and predictable to anyone who has repeated Reich's bion experiments. A significant item in those experiments is iron filings and Reich shows several photographs of iron-filing cultures in *The Bion Experiments*. One item illustrated by Reich that Brown does not appear to have seen is the protruding, moving 'fingers' that Reich called *plasmoids*.[27] These protrude from the surface of an iron particle and move exactly like a waggling finger seen from the side.

Then we come to the *Additional Remarks*. To me this looks like a retraction of Brown's claim that these active molecules are some primary, transitional form of life, as claimed by two earlier workers in the field of spontaneous generation and evolution, De Buffon and Turberville Needham. He denies that he has ever claimed that these particles were alive. Now, he does not say that in so many words, but in his original paper he does seem to imply that they are a step towards life, in the same way that Reich claims that the bions are a transitional form between the non-living and the living. An actual perusal of his 'slips' in the botanical library of the Natural History Museum in London shows that he observed that the movement of at least some of his molecules was pulsatory or contractile in character. In one of these notes he clearly and undeniably uses the word *vermicular* of this motion[28] and he repeats that word in his article. This word means worm-like and so must be referring to some pulsatory process that he had seen. I challenge any reader or scientist to claim that Brownian motion has anything to do with vermicular or pulsatory motion. It appears from a close reading of Brown's own notes (admittedly a daunting task, which one cannot blame people for flinching from), that he was observing both Brownian motion and bionous motility, and that if later scientists had been really persistent in finding out exactly what Brown had observed, they would have realised that there are two types of motion here and that one of them is anything but Brownian motion.

Why should Brown have changed his mind? At this time the controversy about vitalism and spontaneous generation was nasty and severely partisan.[29] Support for spontaneous generation was a materialist, French position, arousing enormous political hostility. (See chapter 4.) People still take sides aggressively and intolerantly even now, when the question is aired. It is quite possible, that someone in the nascent scientific community could have drawn Brown's attention to what they considered was a very risky proposition and warned him that he might be in trouble if he stuck to such a view. It is also possible that someone influential had just suggested that he was wrong and this gave him cold feet. Darwin says that Brown was morbidly afraid of making a mistake.[30] He was also dependent on the good opinion of his superiors. Science in his day was dominated by the rich, financially independent aristocrats of London society. Brown was an exception in this world, a man with no private income who needed a job to survive. He does not appear to have been the sort of man who would have publicly stood his ground in the face of opposition, though I admit that this is sheer supposition. Mabberley's biography is a botanist's biography and gives us very little feel for Brown as a man.

We cannot blame Professor Mabberley for that. It is quite possible that we can no longer get a picture of Brown as an individual. I wondered how I could get more of a feeling for him and how he might have reacted if criticised. I discovered that his diary, written during the voyage to and round Australia under Captain Flinders on the *Investigator* from 1801-1805, when he was the naturalist on board, has been published in Australia. Might this give me some picture of the inner man? I realised this diary did not cover the years in question, when he made his discovery of his active molecules, but I presumed that the diary of any period in Brown's life would give us a fuller picture of what he was like as a human being.

This diary[31] is published with massively detailed notes and allows us to follow Brown's footsteps to the exact places that he visited. When I obtained it I optimistically thought it would answer my questions about Brown as a man. This proved to be a complete illusion. It is an amazingly flat, prosaic document and gives nothing away about his inner feelings, if he had any. It is a prosaic list of where he went and the plants and other specimens that he collected. Frustrating situations, wonderful discoveries, danger, as recorded in the notes, do not inspire any exclamations or outbursts of feeling at all. If a man wants to write about himself for five years without revealing a word about himself, there is nothing we can do about it. Someone who kept themselves to themselves to such an extreme in this way, even in their own diary, would have been an easy person to intimidate, I think. Brown did not, apparently, have a strong sense of self.

Brown's persona, or its complete absence in this diary, must be pathological. Who could travel to the extreme, unknown ends of the world, as one of the first handful of European naturalists lucky enough to visit such parts, see impressive, unique things seen by few, if any, Europeans by that time, experience near-shipwreck, assaults by native peoples, and document nothing but the plain facts of the plants he collected and the physical features of the land that he saw? Only, it seems, Robert Brown!

After this reading we are none the wiser about Brown as a human being. But this massive censoring of his own feelings and reactions speaks volumes, if only negative ones. He was clearly an extremely guarded person who hardly revealed himself to anyone, perhaps not even to himself.

Brown must be the best, the most observant, microscopist that there has ever been. His two discoveries, made with a simple microscope with a magnification of about 400x, are a great achievement. Schleiden (1804-1881), the pioneer of cell theory in biology, acknowledged this. Discussing the merits of the simple and compound microscopes, he pointed out that all recent major discoveries in biology had been made with the compound microscope, '...with the exception of those of Robert Brown (a fine man, who should not be cited here, because he is completely in a class of his own, and his equal will hardly be seen again)…'[32] It may be difficult to see something minute and unexpected, but it is a much greater achievement and far more difficult to see such a thing for the first time ever.

---

[1] Mabberley D J (1985):; Jupiter Botanicus, Cremer, Brainschweig/BM(NH), London.

[2] ibid; page 16.

[3] ibid; chpaters IV-VII.

[4] Lyte C (1980); Joseph Banks, David and Charles, Newton Abbot.

[5] Darwin C (1839); The Voyage of the Beagle, re-published (1989) in Penguin Classics, Penguin, London.

[6] Mabberley D J; op cit, chapter IX.

[7] ibid; chapter XV, The Museum and the Molecules.

[8] ibid; chapter XVI, The Nucleus.

[9] BEOL, pages 57-58.

[10] Brown R; Molecules, page 119, this volume.

[11] Fleck L (1981); Genesis and Development of a Scientific Fact, page 45, University of Chicago Press, Chicago and London.

[12] Dobell C (1960); Antony van Leeuwenhoek and His "Little Animals," chapter 1, The First Observations on "Little Animals" (Protozoa and Bacteria) in Waters, Dover, New York.

[13] Bulloch W (1938); The History of Microbiology, page 29, Oxford University Press, Oxford.

[14] ibid: pages 34-36.

[15] ibid; pages 5-7.

[16] Bastian, H C (1872); The Beginnings of Life, MacMillan, London.

[17] BEOL; chapter 1, The Tension-Charge Formula. (See also *Organ Sensation as a Tool of Natural Research* in EGD.)

[18] Milton R (1994); Forbidden Science, chapter 7, Forbidden Fields, Fourth Estate, London.

[19] Brown R op cit, page 118, this volume.

[20] ibid; page 120.

[21] CB; page 239.

[22] Brown R; Molecules, page 128, this volume and also previous reference.

[23] Brown R; op cit, page 119, this volume.

[24] Palm M and Döring D (1997); Neue Untersuchungen zu den Seesandbionen von Wilhelm Reich, in *Nach Reich*, (eds DeMeo J and Senf B); Zweitausendeins, Frankfurt.

[25] Brown R; op cit, page 119.

[26] ibid, page 119.

[27] BEOL; pages 45 and figures 36 and 37.

[28] Brown R (1827 unpublished); manuscript note in slip number 24/239, Botanical Library, Natural History Museum, London, Thanks to library staff Natalie Pope and Armando Mendez for assistance in locating this item.

[29] Strick J (1992, unpublished); Robert Brown and Brownian Movement, Spontaneous Generation and Microscopy in Nineteenth Century England.

[30] Darwin C (1881); Autobiographies page 60, re-published (2002) in Penguin Classics, Penguin Books, London.

[31] Vallance T G et al (eds) (2001); Nature's Investigator: The Diary of Robert Brown in Australia, 1801-1805, Australian Biological Resources Centre, Canberra.

[32] Schleiden M (1845); cited in Gerlach D (2009); page 211, Geschichte der Mikroskopie, Verlag Harri Deutsch, Frankfurt am Main. (My translation.)

Chapter 2 Wilhelm Reich (1897-1957)

The nucleated vesicle, the fundamental form of all organisation, we must regard as the meeting point between the inorganic and organic – the end of the mineral and the beginning of the vegetable and animal kingdom, which thence start in different directions, but in prefect parallelism and analogy. We have already seen that this nucleated vesicle is itself a type of mature and independent being in the infusory animalcules, as well as the starting point of the foetal progress of every higher individual in creation, both animal and vegetable. We have seen that it is a form that electric agency will produce - though not perhaps usher into full life - in albumen, one of those compound elements of animal bodies, of which another (urea) has been made by artificial means. Remembering these things we are drawn to the supposition, that the first step in the creation of life upon this planet was a *chemico-electric operation, by which simple germinal vesicles were produced.*[1] (Robert Chambers, 1844, his italics.)

Before proceeding to an investigation of other properties of the energy vesicles, we must establish whether the blue vesicles develop exclusively from carbon or from other substances as well. If they were to be found exclusively in carbon, the fundamental question concerning the nature of biological energy in non-living matter would be easy to answer. But the problem is complex, because the more substances we examine and subject to swelling, the more the following conclusion is confirmed; *All matter heated to incandescence and made to swell consists of or disintegrates into blue-glimmering vesicles.*[2] (Wilhelm Reich, The Cancer Biopathy, his italics.)

Wilhelm Reich is a far more accessible scientist than Robert Brown. You may be able to find one of his major works on the bookshelves of a large bookshop in this country, *The Function of the Orgasm.*[3] Amazingly, given how much of an un-person Reich is, this book is still in print in this country and has been for the last twenty years or so. Most local libraries will have a volume or two by him in their dusty stacks, though probably not out on the shelves of their lending libraries. In the seventies the therapeutic part of his work was quite fashionable and he became, briefly, something of a name. That must be why so many libraries still have copies of his better-known works. My own local library service, Lancashire County Council, has all his major works in their reserve stock. However, interest in his work quickly waned, and his books are now being dumped by academic libraries. Many of the hundreds of copies of his books available second-hand are from university libraries. I have bought ex-library copies of his books for C O R E's library. They have often not been borrowed by a single reader.

Reich studied medicine at Vienna University immediately after World War I, in which he had served as an army officer. His parents died while he was still an adolescent and the family farm in the eastern reaches of the Austro-Hungarian Empire had vanished with that empire. He financed his studies by his own efforts, tutoring other students in basic subjects.[4] As a former soldier he was allowed to complete the course in four years instead of the usual six. Even before graduating, he became involved in psychoanalysis and on graduating immediately started work as an analyst. He was obviously very gifted and energetic and quickly became an important figure in the world of psychoanalysis.[5] At first Freud thought highly of him, but disapproved of his strong sexual and political interests.

Reich was particularly interested in what he called the quantitative aspect of emotional life and *libido*.[6] What was libido? Why did its level vary so widely from person to person and even within the same person according to inner or external conditions? He was always drawn to the concrete, the definite, the graspable, and was an outsider, even within the world of psychoanalysis. He was continually striving to make psychoanalytic technique more scientific, reliable and consistent.[7] Few other analysts seem to have shared this passion, though Freud himself was at first keen to build a scientific foundation for psychoanalysis.

To begin with *libido* was just a word for sexuality and emotional excitation. Reich observed that this excitation could be discharged in the sexual embrace, so that there was no more excitation left. It then took time to build up again. Various factors interfered with the individual's capacity to discharge this excitation. He wrote up his observations in his first important book – *Die Funktion des Orgasmus*, which was published in 1927.[8] Although he had made no scientific investigations at that point, the book smells strongly of his future interests and his strong scientific bent.

As he investigated the factors that interfered with the capacity for complete discharge of the libido in the sexual embrace, he expanded the scope of his analytic work and developed a new form of therapy, *character analysis*, which was still only verbal, in which he focused attention on the patient's character defences, rigid attitudes that bound libido, or psychic energy, as he was beginning to call it.[9] He felt that his therapeutic interpretations were bouncing off something hard and impenetrable. He realised that the energy was bound within chronic muscular tension.[10] He had discovered the muscular armouring. When he started to work on the physical tensions themselves to release them, therapy progressed much faster and more deeply. He also worked to remobilise the patient's breathing, which was inhibited by the armouring.[11] As patients became able to breathe more fully and their muscular tensions gave way, they would often report to Reich strong sensations of something moving within their bodies.[12] This sensation was always described as if it were water, something flowing in waves, streamings. (Readers who are not too armoured will recognise descriptions of their own sensations at times of great happiness, ecstasy, or joy.)

As he developed his therapeutic techniques, Reich wondered what the 'something moving' was and whether the sensations of expansion and contraction that his patients reported and the accompanying movements which he observed were functions common to nature in general.[13] He was already seeing common factors in, for example, human breathing and the pulsatory movements of jellyfish.[14] At first he planned to set up an aquarium so that he could observe common marine invertebrates such as jellyfish, starfish, and echinoderms. He changed his plans and decided to study micro-organisms, amoebae.[15]

He received an amoeba culture from Oslo University and set to examining them. When he asked where they came from he was given the usual explanation - from spores or cysts attached to the grass. (Amoebae are cultured in a grass infusion.) He did not accept this explanation, (still the one given in textbooks to this day), and examined the culture more carefully. He also examined grass and the water that he had used to wash the grass in. He was unable to find spores attached to the grass or in the water that the grass had been washed in.[16]

He next observed what happened to grass when it was soaked under water. He devised an ingenious but simple technique that allowed him (and which allows us today) to observe the process that grass undergoes when it swells and starts to disintegrate under water. (If you have been wondering what all this has to do with Robert Brown, we are now getting there, to the point where the two areas of research meet.) Reich saw that grass under water swells slowly, starts to move gently, and breaks down into tiny, very motile, highly charged vesicles. These sometimes form clumps within a membrane on the edge of the grass-blade and eventually break away as independent organisms.[17]

Reich realised, unlike Brown, that the tiny vesicles that he saw appearing were the products of a process, *bionous disintegration*. They were not in the grass to start with: they developed from it.[18]

Reich's scientific reaction to his discovery was similar to Brown's. After testing other vegetable tissues, grass and moss, he moved on to soil, humus, charcoal, and coal dust. Just as did Brown, the next step that Reich took was to see what happened when he did the same tests with completely inanimate materials, things that had never been alive – sand and iron filings. They also produced the bions that he had discovered.[19] The obvious objections to his findings were those of contamination and our old friend, Brownian motion. To be sure that his bions were not cocci, spherical micro-organisms, he repeated the control experiment that had occurred to Brown. He heated his dry ground materials to red heat.[20] Surely there is no organism that can possibly survive such a heat? (There are one or two micro-organisms that can survive autoclaving, notably *bacillus subtilis*.) How hot is red heat? A technical dictionary in C O R E's reference library says: 'as judged visually, between 500° and 1000°C.'[21]

No organism can survive such temperatures. If we heat any material containing organic traces, this heat burns them off immediately. Reich found that this treatment hastened the process of bionous disintegration and produced more bions more quickly.[22] I shall point out here, just in case readers are thinking that this heat treatment is difficult and the sort of thing one can only do in a proper laboratory, that we can do this with a spatula and the flame of a gas cooker. If you have no spatula, you can use the blade of a knife. And if you have no gas cooker, you can use the flame of a blowlamp or a methylated spirit burner. I live in an all-electric flat and use such a burner for my experiments. When I lived in a house with a gas-cooker in the kitchen, I heated samples in the cooker flame.

So …if Reich is right and he has found something new to science, something that no-one else has seen and described … how do bions move and how can we be sure that the movement that we see in them is not simple Brownian motion or contamination?

I do not know of any scientist who has actually repeated Reich's experiments in detail, not confirmed Reich's findings, and reported his or her results publicly. As far as I know from my reading, all the criticisms have been in the 'It must be contamination…it must be Brownian motion' vein. In other words, the critics have not repeated the experiments and have just plucked the first excuse out of the air that they can use to dismiss Reich's findings. A S Neill, the Scottish libertarian educator and Reich's great friend, quotes these objections from well-known British scientists frequently in the collection of their letters published in 1985.[23]

If the bionous motion that we see is really Brownian motion, how do such critics explain the fact that we get what they would have is Brownian motion in an acidic solution of limestone in vinegar at pH 4.0 and no Brownian motion with the same materials in water at a pH of 12.0 or above? I have never come across any suggestions by anyone that Brownian motion has something to do with pH values. It is always exclaimed as purely mechanical motion caused by the kinetic energy of the molecules of the surrounding fluid in which the particles are immersed.[24] We can also carry out a simple control experiment of the grass-infusion preparation. This is carried out according to Reich's protocols with the slide open to the air and so is open to the accusation that the organisms that we find on the slide have got there from the air. (We do in fact eventually see organisms from the air in the grass infusion preparation. They always seem to be the same rods and occasional cocci and vibrios. Reich includes in his reports a photograph of the common forms[25] and all these years later the same forms appear quite predictably and regularly.) It is not too difficult an experiment to prepare a sterile test-tube of boiled water and a few chopped up blades of grass and to autoclave this tube so that we know that it is going to be more or less sterile and should not therefore produce any motile forms, as does the open grass-infusion preparation. We can seal the tube with a *Subaseal* stopper. This allows us to withdraw a small sample of the fluid in the tube without opening it. (See chapter 4 for more details.) To be even more certain that we have eliminated all contamination, we boil the preparation for 30 minutes and incubate it at 30°C for a day in between boilings. We do this three or more times and then autoclave it. We still see almost all the same forms as Reich and others report in the grass-infusion experiment.[26]

The easiest way for the student new to orgonomy and the microscope to see bionous movement and bions themselves is to make a preparation in a test-tube from some commonly available material that produces bions quickly and reliably, such as clay, coal-dust, most types of ground stone, ground charcoal, iron filings, sand, and so on. It is very interesting to test all sorts of materials for bionous growth, once you have got started and are familiar with the forms and methods, but I do recommend the easy materials to start with. When I had repeated Reich's original experiments and found my feet in the field, I became a bion bore and ground and tested every conceivable item I could find, to see if it produced bions. Less likely things that I tested were ground orange peel, ground potato starch, granite and gneiss, (which I picked up on holiday on the Isle of Mull), artificially made sand from ground up shells, crab-shells, and seaside detritus, and so on. I obtained samples of common materials such as slate and brick from builders' skips. For the practical details of these investigations, please see chapter 7.

It is important that students carefully observe the ways in which bions move, so that they can meaningfully distinguish between bionous motility and Brownian motion. It should also be a basic part of any student orgonomist's education to examine a variety of suspensions under their microscope to find out exactly what true Brownian motion looks like. (If you are lucky enough to have access to very high magnification, above about 3000x, you can even see the blue glimmerings of the orgone energy charging the bions' movement and the pulsation within the bions themselves, though few microscopes are equipped with magnification to this level. Such very high magnification is always supposed to be 'empty', that is, to show no more detail or definition because of the laws of light that govern the possibilities of the microscope. We are looking for and at movement, not definition or structure, and so this high magnification is meaningful and relevant to the orgonomic microscopist.)

In a bion culture we see particles of a similar size, some of which are moving and some of which are not. In a suspension showing Brownian motion all the particles of a similar size will be moving in more or less the same way, moving from side to side in all directions and also up and down in the fluid, ie, away from the viewer's eyes and towards them, so that a particle keeps moving into and out of focus while we observe it. If we prepare a bion culture from a material that gives a borderline pH of around 12.0 we may even be lucky enough to see bions starting to move and even 'dying', that is, starting to move frantically, far more vigorously than a bion usually moves, and then suddenly ceasing all movement.

It is vital at this stage to look very carefully indeed at your preparation and not just to dismiss what you see as Brownian motion. (You will be tempted to do this, if you are sceptical towards my arguments.) If you are looking at bions, you should, after a while spent scanning your slide, find here and there pairs of bions dancing around each other. This is the attraction and repulsion function first observed and described by Reich in *The Cancer Biopathy*.[27] You may also observe pairs of bions linked by what Reich called a *radiating bridge*. You will see larger bionous particles of your base material, much too large for Brownian motion, that are wobbling gently and even rolling over themselves slowly. You may also find very small particles that are rotating furiously. In some preparations you will see agglomerations of the bions. These look rather like small insects and often bend and twitch. Brown described this motion in his article.[28] In some borderline cultures the bions that you can see will stop moving quite shortly after they have come into existence. (Also any moving particles will stop moving if they sink down onto the slide surface.) Reich's experimental results were greeted with scorn and dismissed out of hand as the misguided interpretation of either Brownian motion or contamination. This was in the usual format of – that must be… that can only be… No-one went off to their laboratory, carefully repeated the experiments, and found that they obtained different results.

Soon after this period in his life-work Reich emigrated to the United States. He quickly re-established himself and his research in New York and started training new students. This was a very fertile time in his life and he went on obtaining important findings. Although ignored by sceptical mainstream science and medicine, he continued to attract a small number of doctors and scientists who wanted to train in his field. He went on making discoveries and obtaining important findings almost to the end of his life. The story of his persecution and imprisonment can be read in detail in the two biographical works on the history.[29] The FDA claimed to have tested many of Reich's findings and to have disproved them as fraudulent, but when under the later Freedom of Information Act researchers unearthed what exactly they had done, it turned out that they had conducted their investigations carelessly and not taken any of the major precautions that Reich said should be taken to obtain accurate results.[30] They did not investigate the findings of his bion experiments, presumably as these were not part of any medical treatment or claims.

Mainstream science has not been able to steal Reich's discovery and give it another label, as it has, fortuitously, been able to do with Brown's discovery. What luck for science that there was another real phenomenon involved, what is now called Brownian motion. It has just ignored Reich's great discoveries of bionous disintegration, the origin of the bions, and the discovery of the orgone in the hope that they will go away. And they have more or less gone away, simply because if you never mention something, it will eventually disappear from people's awareness. Another tactic indulged in by a few extremists with, it seems, time to spare, is to write deeply hostile, abusive articles about Reich and orgonomy,[31] so that the innocent student of science comes across these and automatically assumes that they are truthful versions of the facts and that there is no need to follow up the original articles or books. Reich is such an un-person now that I do not suppose any scientist under the age of 40 has even heard of him. He very occasionally gets a fleeting reference in book reviews.

---

[1] Chambers R (1844); Vestiges of the Natural History of Creation, pages 204-205, reprinted (1969), Leicester Universty Press, Leicester.

[2] CB; page 20.

[3] FO.

[4] Sharaf M (1983); Fury on Earth, page 55, André Deutsch, London, and PY, page 73.

[5] ibid; part III.

[6] FO: chapter I, Biology and Sexology before Freud.

[7] ibid; chapter III, 3, Founding of the Vienna Seminar for Psychoanalytic Technique.

[8] Reich W (1927); Die Funktion des Orgasmus, Psychoanalytischer Verlag, Vienna, re-published in English as **GTTN**.

[9] FO; chapter V, The Development of the Character-Analytic Technique.

[10] ibid; page 270.

[11] FO; chapter VIII, 4, The Establishment of Natural Respiration.

[12] ibid, page 271.

[13] BISA; chapter 3, The Bioelectrical Function of Sexuality and Anxiety.

[14] FO; chapter VIII, 4, The Establishment of Natural Respiration.

[15] BP; page 63, footnote.

[16] CB; chapter III, 1, The Absurdities of the Air-Germ Theory; and in particular pages 77-78

[17] BEOL; pages 31-37 and figures 17-25

[18] op cit; pages 26-31, Vesicle Formation in Swelling Blades of Grass.

[19] CA; chapter II, 1, The Vesicular Disintegration of Swelling Matter, (PA Bions).

[20] Brown R; op cit, page 121, this volume.

[21] Walker P M B (ed) (1995); Larousse Dictionary of Science and Technology, *red heat*, page 920, Larousse, Edinburgh.

[22] BEOL; page 49, and CB, page 18.

[23] Placzek B R (ed) (1982); Record of a Friendship; The Correspondence of Wilhelm Reich and A. S. Neill, page 106, Gollancz, London.

[24] Illingworth V (ed) (1990); Penguin Dictionary of Physics, page 46, *Brownian movement or motion*, Penguin Books, London.

[25] CB; figure 33.

[26] Jones P (2007); Three Experiments with the Microscope for the Amateur Orgonomist, C O R E, Preston.

[27] CB; pages 52-53.

[28] Brown R (1828); Molecules, page 121, this volume.

[29] Sharaf M (1983); op cit, chapters 28-32.

Greenfield J (1974); Wilhelm Reich vs the USA, Norton, New York.

[30] ibid, Appendices 8 and 9.

[31] Gardner M (1957); Facts and Fallacies in the Name of Science, Dover, New York.

Chapter 3 Dismissing Reich

> When the works of these charismatic figures are carefully perused, they often make little or no sense, as for instance the balderdash of Wilhelm Reich's orgonomy.[1] (The History of Psychoanalysis, page 78.)

> Haldane dismissed your *Bione* as "too unorthodox" which is about the silliest thing a scientist could say.[2] (Record of a Friendship, The Correspondence of Wilhelm Reich and A. S. Neill.)

The genuinely open-minded and curious reader coming across this information for the first time may be wondering where it has been all this time and why they have not come across it before. The answer is that there is a quiet conspiracy to either avoid completely the facts revealed here or, if they threaten to emerge, to ridicule and traduce them so severely that no scientist, reviewer, or historian will ever bother to take them seriously or to include them without bias in any text-book or reference work. (We have seen a massive dose of this ridicule and contempt recently (2011) in the UK in the form of reviews of a malicious and ignorant book, obviously written by people who know nothing of orgonomy.) The non-scientist will find it hard to believe that some professional scientists in the US contribute articles to orgonomic journals under pseudonyms for fear of what public knowledge of their involvement in orgonomy will do to their academic reputation. Such knowledge would ruin their career. (See for example author's details in issue No 5 of *Pulse of the Planet*.[3]) I do not know of any British scientists with a commitment to orgonomy who do this. As far as I know there are no such scientists in this country. Orgonomy is just not on the map here, not even amongst scientists and doctors with interests in alternative models of science and medicine.

I have used the expression conspiracy above, but I do not at all believe that scientists and heads of colleges and research institutions have ganged up and decided to boycott anyone who feels positive about orgonomy. It is simply an unspoken assumption that affects any theory, suggestion, or project that assumes that there may possibly be a life force in nature. If one thinks that there may indeed be such a force or proposes any theory or research project that assumes the possibility of such a force, one finds oneself immediately beyond the pale. One cannot get any serious discussion going on any such topic or project and one certainly cannot get any serious research published in a mainstream scientific journal. Because of this boycott, all orgonomic writings are published in privately printed journals financed by orgonomic scientists and research workers themselves. The circulation of these journals is tiny and one does not come across them by accident.

How has this unbelievable situation come about? It began in Oslo when Reich started his pioneering work on the bions and cancer, work that we are connecting with in this book. As soon as this work became public knowledge, he was attacked as a fraud and criticised for claims that he had in fact never made.[4] This sort of attack continues to this day, on the rare occasions when people pay Reich and his work any attention. He was represented in a cartoon as a high priest revealing the secrets of life to worshipful acolytes, because he claimed that he had observed the first steps between non-life and life.[5] Note that no-one repeated his experiments and claimed that the results obtained did not confirm Reich's own claims. They just abused him and dismissed him verbally with the two classic objections – Brownian motion and contamination.[6] We shall come to Brownian motion later, when we repeat Brown's own experiments. What about contamination? How pertinent to his work is this criticism? We need to make a short excursion into the history of microbiology and theories of infection in the nineteenth century before we can judge this criticism and Reich's reaction to it and his control precautions.

Reich was not the first scientist to claim that he had seen living or transitional forms originating in sterile preparations. Many other scientists before him thought they had seen similar forms originate from heated sterile preparations and this started a rather fruitless debate that carried on for decades until it was presumed to have been solved once and for all by the results of Pasteur's famous experiment with his swan-necked flasks.[7] Pictures of this experiment appear in almost all undergraduate textbooks of biology and microbiology. Pasteur claimed that these preparations open to the air, but closed to the access of air-germs, demonstrated conclusively that living forms did not originate in cultures when they were closed to contamination from the air.[8] The preparations in question were all fluids containing no solid matter.

Before Pasteur the English catholic priest Turberville Needham (1713-1781) claimed he had found living forms in preparations of meat that were sealed and sterilised by heat.[9] Spallanzani (1729-1799) claimed to have demonstrated that Needham's sterile precautions were inadequate and that his containers allowed in air.[10] This demonstration did not prevent later researchers, notably Félix Pouchet (1800-1872)[11] and H C Bastian (1837-1915)[12], from claiming that they had also found living forms in sterile preparations. The arguments over these claims are boring and fruitless and revolve round the temperature at which all forms of micro-life are killed or not.[13] It is not an oversimplification to say that whenever a researcher claimed to have found these living forms his critics automatically retorted that he had not sterilised them at a high enough temperature. This temperature drifted slowly higher and higher as forms were discovered that could apparently survive temperatures as high as 120°C.[14] Nowadays, in hospitals and laboratories, the generally accepted temperature and time for autoclaving surgical instruments is half an hour at 120°C. As we will see below in chapter 7, we still get bions in preparations sterilised for longer than this period and at a slightly higher temperature. (And we also get bions from solid materials that have been heated to red heat.)

Needless to say, those who were ideologically and psychologically disposed to be against spontaneous generation, as it was often called, made this objection, while those who were psychologically in favour of spontaneous generation discounted these objections. Neither party was strictly scientific, but interpreted results according to their own bias.[15] This situation has not changed much since the middle of the nineteenth century. Reich was, therefore, entering a dangerous area and stirring up a hornet's nest with his bion research. Well-known professors wrote hostile articles in the daily press criticising Reich *ad hominem* and ridiculing his research. Though some of them claimed to have looked at his preparations, no-one was interested enough to repeat the experiments themselves. They simply dismissed the apparently living motile forms as Brownian motion or contamination. Reich sent some preparations to his friend A S Neill, the Scottish educator, who knew several well-known scientists of the day, amongst them J B S Haldane and J D Bernal. Their responses were of the usual sort: the living forms must be Brownian motion or contamination.[16] Bastian, too, had to put up with this 'it must be…' response.[17]

Ordinary working scientists hear these claims, assume they are true, and repeat them in the same way that they mindlessly repeat the neutered version of Brown's discovery. We come across entries about Reich and his work in respected works of reference that are complete rubbish, presumably based on this sort of scientific gossip.[18] No reader, new to Reich and orgonomy, would ever guess that Reich left behind a large amount of experimental evidence and records that allow us to repeat his experiments and to test his claims. Reich's claims and orgonomy in general are presented as if he made it all up, like some nutter's concocted philosophy. There is frequent reference to his orgone 'theory', rarely, if ever, any to his experiments.[19] This continues to this day, in spite of the fact that the information that we need to test Reich's claims is out in the world and available to anyone who wishes to test them. (Of course, orthodox scientists are not interested in testing Reich's claims, as they might discover more than they bargained for, if they did test them.)

There is also the relentless, repetitive drip-drip of the conventional models of biology and medicine. Findings and experiments allegedly proving these models and theories are repeated in almost every textbook one cares to look in. These are repeated in a bland, unquestioning way with no suggestion that there might be any differing interpretation of the facts or experiments and theories. Differing interpretations never get a mention in mainstream works. If we study the history of spontaneous generation and vitalism in the nineteenth century, it is clear that nothing is really proved or disproved. It is simply that the faction affirming the occurrence of spontaneous generation lost a political battle and once they had been defeated in this way, no-one new to science dared embrace these theories and models, and the old supporters died off one by one until there were none left. When H C Bastian died in 1915 his obituaries expressed relief that the last, wearisome, obstinate advocate of a dead-and-buried doctrine, the last flat-earther, had finally died.[20] If we think of this history as a conspiracy in a family not to ever mention an embarrassing episode in the family's history, then Bastian was the last living witness to the crime. He was now dead and everyone could at last sleep peacefully at night. There was no-one left who could spill the beans.

The repetition of these dogmas is so low-key and apparently unmalicious that the innocent student of biology can be forgiven for accepting them without demur. While writing this I have come across a good example of this drip-drip in an excellent book on freshwater life in this country. The writer of this book has a sensitive feeling for life, writes beautifully, simply, and clearly. It just cannot be that he intends his words maliciously. He has heard these explanations so often that he cannot believe anything else, in the same way that someone who has been brought up in a religion just cannot believe that what they have been taught is made up by other human beings and only exists in the minds of those who have been similarly brainwashed. I repeat his words here so that readers will have an impression of what I am talking about. (Even a conventional authority on the distribution of protozoan cysts could apparently find only 2.5 per cubic metre at ground level on a river bank and concluded that this was far too low a density to account for their world wide presence and similarity and that therefore there must be other dispersal paths.[21] He would not have allowed the possibility of other routes of origin.)

The protozoa and other simple organisms which were considered to be related to them were formerly classified in one heterogeneous group, the *Infusoria*, because they appear in a "hay infusion"; that is, water in which a small quantity of hay or other dried vegetable matter is left for a few days.

> They nearly all possess the ability to withstand drying up by surrounding themselves with a resistant covering, and their desiccated forms, left perhaps on aquatic vegetation by the lowering of the water level, or maybe stranded on the mud at the edge of a pond, are blown about in the air in such numbers that almost any plant will be covered with them. When an "infusion" is made, they burst into activity again as if nothing had happened. (J Clegg, *The Freshwater Life of the British Isles*, Warne, Third edition, 1965.[22])

Another strand of criticism is the rumour, which first arose during Reich's time in Norway, that he went mad. When he is alleged to have gone mad depends on the point at which the critic felt unable to follow his work any further. For one critic it is Norway in 1936, if not earlier, for others it is when he arrives in the US, and so on, on and on.[23] Now if someone is mad, we do not need to even consider them as a real, relevant thinker or scientist, do we? Reich is in good company in this group of alleged madmen – that of Semmelweiss, William Blake, and Giordano Bruno. Even amongst admirers Reich's work is often misunderstood and bowdlerised. An article on him in the magazine *Nexus* some years ago claimed that he had been persecuted by the oil industry because his discovery of orgone energy promised free energy and therefore was a threat to it. This is complete fantasy.

As far as I can see, Reich does not formally cite Brown's *Active Molecules* paper anywhere, but he was clearly familiar with it. In the chapter on his research methodology at the end of *The Bion Experiments* he writes that Brown originally thought that the motion which he had observed was biological in nature.

> Brown believed he had discovered a living phenomenon. Physicists, however, explain the phenomenon as a purely physical manifestation brought about by molecular motion and, contrary to the view of its discoverer, Brownian motion is now established as a concept with which one explains *any* microscopic movement that is not clearly and obviously associated with living matter. It is not pleasant to hear an expression used over and over again when one is demonstrating the existence of other totally unrelated phenomena. One physicist admitted to me that he was teaching Brownian

movement to junior high school students although he had never seen it himself.[24]

It seems that Reich was familiar with Brown's original 1828 paper, even though he does not include it in the bibliography of *The Bion Experiments*.

As Reich wrote above, the greatest fraud of all is science's use of the discovery of Brownian motion to ridicule Reich and other workers, too, when in fact the original *Active Molecules* paper clearly describes bions and bionous motility! If you find this claim extravagant, I urge you to read Brown's original paper, (Appendix 1), and to repeat his original experiments, (Chapter 7). If you are not at present in a position to repeat these experiments, please think of buying or borrowing a microscope and repeating them yourself. As you will see, if you read chapter 7 carefully, this is not as hare-brained a project as it might appear at first sight.

---

1 Fine R (1990); The History of Psychoanalysis, page 78, Continuum Publishing Company, New York.

2 Plazcek B R (ed) (1982); Record of a Friendship: The Correspondence of Wilhelm Reich and A S Neill, page 106, Gollancz, London.

3 De Meo J (ed) (2002); *Pulse of the Planet*, No 5, page 88, Orgone Biophysical Research Laboratory, Ashland, Oregon.

4 Sharaf M (1983); Fury on Earth, pages 229-233, André Deutsch, London

5 BP; cartoon included in illustrations between pages 102-103.

6 CB; chapter II, 2, The Question of "Brownian Movement", and chapter III, 1, Absurdities of the Air-Germ theory.

7 Madigan M T et al (1997); Brock Biology of Microorganisms, page 22, Prentice-Hall, Upper Saddle River New Jersey.

8 Farley J (1977); The Spontaneous Generation Controversy from Déscartes to Oparin, chapter 6, The French Coup de Grâce, Johns Hopkins University Press, Baltimore, Maryland.

9 Needham T (1748); A Summary of some late Observations upon the Generation, Composition, and Decomposition of Animal and Vegetable Substances, in *Phil. Trans. Roy. Soc.* Vol 45, 1748, pages 615-667, London.

[10] Spallanzani L (1803): Observations and Experiments on the Animalcula of Infusions, chapter I, Creech and Constable Ediburgh, in Strick J (ed) (2004); The Origin of Life Debate: Molecules, Cells and Generation, volume I, Thoemmes-Continuum, Bristol.

[11] Pouchet F A (1859); Hétérogénie, ou Traité de la Génération Spontanée, Baillière, Paris.

[12] Bastian H C (1872); The Beginnings of Life, MacMillan, London.

[13] Farley J; op cit, passim.

[14] ibid, passim.

[12] Strick J (2000); Sparks of Life, passim, Harvard University Press, Cambridge, Massachusetts.

[16] Plazcek B R (1982); op cit, page 23, Gollancz, London.

[17] Strick J (2000); op cit, page 89, Harvard University Press, Cambridge, Massachusetts.

[18] See for example the entries under *Wilhelm Reich* and *orgone therapy* in the Oxford Dictionary of Psychology (second edition 2006). Thanks to my prompting, the entries in the third edition are more accurate.

[19] Honderich E (ed) (1995); The Oxford Companion to Philosophy, page 753, Oxford University Press, Oxford.

[20] Anon (1915); The Death of Dr. Bastian, Obituary in *The Times*, November 1915, London.

[21] Puschkarew B M (1913): Über die Verbreitung der Süsswassprotozoen durch die Luft, *Arch Protistentk.*, 28, pages 323-362. Cited in Gregory P H (1961); The Microbiology of the Atmosphere, Leonard Hill [Books] Limited, London.

[22] Clegg, J (1965); The Freshwater Life of the British Isles, page 91-92, Warne, London.

[23] Sharaf M; op cit, passim.

[24] BEOL; page 149.

Chapter 4 Rebellion from Below

The Social, Personal, and Political Background to Brown's Retraction

An unbiassed reading of Brown's first paper and his later *Additional Remarks* makes it clear that he had changed his mind (or had it changed for him) between 1828 and 1829. How could this have come about? There are two strong influences at work here – those within Brown, his own emotional character, and those without, the surrounding social and political ambience. Brown's persona on its own is enough to account for his change of mind.

We already know that Brown had a morbid fear of making a mistake and appearing to be in the wrong. Mabberley says of him in his entry in the DNB that he was -

> frustratingly cautious in publishing...[his scientific findings]...wishing never to make a mistake in print. His procrastination may have been due to the perfectionism of an obsessional personality.[1]

As we have already seen, Darwin made the same comment about him.[2] This raises the question of why he apparently hurried into print with unclear findings and had this paper printed privately, when normally he was known for being a terrible procrastinator. This suggests an unusual (for Brown) level of excitement and interest in this particular project, as if he felt he had discovered something of more than common importance. After all, there is a great difference between the discovery, collecting, and naming of a new species of plant and discovering a new principle of nature, a process by which apparently living proto-organisms come into being. Anyone could do the first. Only an original, as great an observer and microscopist as Brown, could do the second.

Some of Brown's written works remained unfinished. His planned *Prodromus*, an introductory treatise on the plants of Australia, was never completed, in spite of proddings by Sir Joseph Banks.[3] According to Mabberley, colleagues found his reluctance to complete planned projects extremely frustrating.[4]

In view of this well known fear of making a mistake, and Brown's severely reserved and unassertive personality, it would have been very easy for someone with power and influence to have raised doubts in Brown's mind and to have made him feel much less certain of his findings. It is very difficult to imagine Brown standing up publicly to criticism and the tide of scientific fashion, even possible warnings (see below), and insisting on the rightness of his findings. He was no scientific hero. It is rather strange that he published no further research in this area, apparently so important to him, and made no further efforts, (at least not publicly), to clarify his findings. He was an extraordinarily good observer and microscopist and must surely have been capable of doing that. Ford's recent investigations leave us in no doubt of the accuracy of Brown's observations.[5] Another possible explanation of Brown's behaviour, which fits his procrastinating nature, is that, like many of us, he was going to get round to it later, and never did, because he had so many other things to do. He may have been unconsciously deterred by his awareness of the enormous controversy of the questions he had touched on. Poor Brown - the first researcher to touch experimentally upon the origin of life. Little did he know what a can of worms he was poking into.

I think all the above alone can explain Brown's change of heart and mind. However, if we go outside Brown and have a look at the surrounding political and social ambience of his day, in fact, of the first decades of his adult life, we see that there were huge pressures around him that would have been hostile to his first interpretation of his findings, that his active molecules were…

> in reality the supposed constituent or elementary Molecules of organic bodies, first so considered by Buffon and Needham, …

By making a connection with these two men, Brown was associating himself with the French tradition of a belief in spontaneous generation. But this was tainted with materialism and atheism, so much so that when Needham wrote his 1748 paper for the Royal Society, summarising his work, he felt the need to explain to readers and the gentlemen of the society that, though these theories (of spontaneous generation and a productive force in nature) had in the past been associated with materialism, he was by no means a materialist and was in fact a thorough-going believer in God. (After all, he was a Catholic priest.)* I

---

* Today a scientist would feel they had to disavow any vitalistic beliefs, if they were publishing something controversial which might be interpreted as 'vitalistic.'

noticed this defence of Needham's before he has been attacked myself and Strick makes exactly the same point in his unpublished essay on this background.[6]

> This Exposition, Sir, of my Sentiments, I thought might be necessary; not that I imagined that either you or any of the Gentleman of the learned *Society* in which you preside, would think of my Principles any way tending to Materialism, from which no one can be more distant or averse than myself; for I well knew that I had nothing to apprehend of Persons of so much judgment and Discernment, and would not but clearly see, that there is really no connection between those Principles rightly explained, and the doctrine of the materialists: But I was willing to guard against the Misapprehension of others less acquainted with Matters of this sort, and into whose Hands this Paper might come, and have therefore taken these precautions.[7]

To understand the pressures and influences that could have been so decisive in the private and public interpretation of his findings we need to make a short excursion into English history of this period.

It is difficult to define the important influences on Brown during the first quarter of the nineteenth century, so nebulous and invisible, almost 200 years after the events. We are bumping up against the huge difference between a historic battle or building or event recorded in books and paintings and a social ambience that has disappeared and of which we can find no traces whatever without trawling through long-buried archives and history books.

Most people leaving school with any knowledge at all of British history have some solid dates and landmarks in their minds, even if they do not read another page of history for the rest of their lives. Most educated people will immediately recognise 1066, 1588, 1642-1660, 1805, 1914-1918 and 1939-1945. We all know of The Battle of Hastings, The Spanish Armada, The Civil War and Commonwealth, the Battle of Trafalgar, the Crimean War, Queen Victoria's reign and the two World Wars. These are solid identifiable items in our history. Even now we can go and 'see' these

---

I have come across one such disavowal during my research into the origin of life, a few lines in the introduction to Maturana and Varela's *Autopoiesis and Cognition: the Realisation of the Living*, 1991, Springer, New York.

events, or if not see them, see plenty on the ground to demonstrate that they actually occurred or existed at one time or another. We can go and visit a Civil War battlefield, see Queen Victoria's train in a museum, visit Nelson's flagship, *Victory*, and see relics from the World Wars in museums. There is no doubt that these items or events were real.

Brown and his contemporaries lived through something that had just as much influence on British history and society as any of the events mentioned above, but this something is now long forgotten, except by historians, and completely invisible to lay-people. Even professional historians find it difficult to pin down its events and participants' opinions, so nebulous is it and so vague is its meagre remaining evidence. This something is the long period of social tension and reaction in Britain between the French revolution in 1789 and the first Reform Act of 1832.

Readers may be bridling at the thought of any connection between this and Brown's experimental work. Science is science and has nothing to do with politics, has it? Well…please read on. Please don't throw the book into the bin out of impatience and disbelief.

It is hard for someone living in Britain now to imagine the tension, paranoia, even hysteria, that prevailed much of the time during this period and how a morbid fear of rebellion from below governed the attitudes and actions of the governing class. Many campaigners and reformers fell foul of the law. There were numbers of executions, and many radicals and revolutionaries were imprisoned or given long sentences of transportation to Australia. As late as 1826 a rising in Lancashire was 'smothered by military terror.'[8] Reading of the trials, arrests, the governement's use of informers and *agents provocateurs*, and the use of the army to control the population of the new industrial areas, one is reminded of the gulag and the 'justice system' of a totalitarian state. During this period the government constructed army barracks in industrial areas for the purpose of controlling the working population. They used 12,000 troops to control the 1812 outbreak of machine-breaking in Yorkshire and Lancashire.[9]

Brown was born in 1773 and would have grown into manhood as this culture of reaction and governmental nervousness was taking root. (Mabberley does not mention any of this important background to Brown's life in his biography.) He disembarked from his Australian voyage at Liverpool in 1805 in the heart of this reaction and its effects.

It is not surprising that the governemnt and the ruling class were so jittery during this period. The war with France was causing major disruption to trade and a rise in food prices and a fall in wages in many industrial areas. Society was undergoing deep social stress and disruption as the industrial revolution started to bite seriously. All this was taking place in a society in which only 10% of the population had the vote in 1793, fewer proportionately than at the beginning of the eighteenth century, thanks to the greatly increased population. (The percentage was falling all the time as the population grew. Prior to the 1832 Reform Act it was about 3 %.[10]) In Scotland then only one percent of the population had the right to vote. The respectable classes, those with status, power and influence, with whom Brown was mixing and working, the likes of Sir Joseph Banks and fellow gentleman-scientists, must have felt they were sitting on a powder keg. *Habeas corpus* was twice repealed in 1794 and 1817, so afraid were the government of rebellion and plotting.

During this period the government was building barracks in industrial areas for the army so they could better police rebellious or revolutionary workers. The concept of a police state did not exist at that time, but surely Britain was not far from being a police state, if you were a worker in the new industrial society. The Combination Acts making trade unions illegal until 1824 were re-introduced in 1825 after a wave of strikes and industrial unrest.[11]

A passage in *The New Oxford History of England* describes the atmosphere and attitudes unforgettably:

> Demonization of the poor as potential revolutionaries had been commonplace since the last two decades of the eighteenth century.
>
> ...such images [referring to caricatures of the poor and the rich in cartoons by Gillray] were symptomatic of polite society's inability to distinguish between poverty and various types of deviance, such as crime, delinquency, lunacy, sexual depravity, and Jacobinism. In their imaginations all these horrors merged into one great phantasmagoria of the mad, bad, dangerous people, an infectious disease threatening to destroy civilization. 'Pauperism we consider nearly as infectious as smallpox,' wrote one official. 'Without constant vigilance it would soon overspread the whole parish.' Like speculation and disease, poverty seemed capable of being spread through personal contact until it engulfed the whole of society. These phobias affected mainstream society for as long as revolution seemed a possibility, i.e. until the mid-century,...[12]

Further, from the same author, almost as asides, so taken for granted is this feeling of the respectable classes about the rest of society, 'Once fear of the mob had receded...'[13] and 'when after 1850 the fear of revolution disappeared'...[14]

A surprising feature of this period is that this division in society between the threatening, rebellious masses and the staid, orderly, reliable and safe governing classes extended into science. This appears to have started with the phobia of the French revolution and its Great Terror. By the seventeen nineties there was already French and therefore radical revolutionary science and conservative British science. The destruction of Joseph Priestley's house, laboratory, and library by a 'patriotic' mob in 1791 reflects this division and hostility to 'revolutionary' French science. Of course Priestley's science was only science, but he had contacts with the great French chemist, Lavoisier and exchanged information with him, as any open-minded scientist of that period or any other period would have done with eminent colleagues.

Spontaneous generation was already branded with the sins of atheistic materialism in France before the revolution. According to Farley, who has written the definitive history of spontaneous generation, the *philosophes*, the supporters of the belief in spontaneous generation, were blamed for 'generating ideas leading to the chaos of the French Revolution.'[15] This was exactly the fear of the respectable classes in Britain, that materialistic ideas and values imported from France would encourage social breakdown and revolution.

Sir Joseph Banks, a landed aristocrat with massive private wealth was clearly part of the governing class and going to be resistant to anything that smacked of revolution. He did everything he could to draw high-society amateurs into science to give it status and influence. (He died in 1820 and so would not have been alive to affect Brown's interpretation of his findings.) Brown was not a moneyed aristocrat and had to earn his own living. He was employed by Banks for the last ten years of Banks's life and associated with him for years. Brown therefore was surrounded by the governing classes and must have been aware of and influenced by their attitudes. Almost all scientists in the late eighteenth and early nineteenth centuries were gentlemen-scientists of independent means, as there were few formal scientific institutions and few salaried jobs as scientists until after 1850, if not later. Even then T H Huxley was still complaining bitterly that a man of science would not earn 'enough income to pay his cab hire.'[16]

When Banks's herbarium was eventually moved to the British Museum in 1827 after his death, Brown joined the handful of regularly employed curators and librarians there who were maintaining the collections.[17]

As we shall see shortly, this conservative, even aristocratic background to Brown's life and work and the social pressures on him must have played a significant part in his attitudes towards his work and the world about him.

Because of the split into French science, radical, godless and materialistic, and British science, religious and conservative, the police-state tendency reached into the actual work of science and medicine. The fate of William Lawrence and his retraction of his radical, materialist opinions in 1820, only seven years before Brown made his discovery of bionous motility and Brownian motion, shows how deep this reach was. Lawrence was, as a student, the protegé of the well-known conservative surgeon, John Abernethy. He was a radical spirit, and, apparently, in his early days, not afraid to speak out and say what he thought. It was made quite clear to him that this just would not do. He was suspended from his post as surgeon at the Bridewell and Bethlehem hospitals in London and to regain it he was forced, by a mixture of moral blackmail and pressure from his employers, to renounce in writing his materialistic leanings and to withdraw his own anatomical text-book from circulation. As well as this hostility from his employers, he was surrounded by a barrage of hostile critical articles. The words *blasphemous*, *seditious*, and *immoral* were thrown at him. His opinions were not being shouted from the rooftops nor being published as cheap broadsheets that any literate person could read. They were first heard as lectures at the Royal College of Surgeons and were then printed as a medical textbook.[18] Yet, apparently, he was a threat to the nation's very fabric.*

Radicals liked his book and published two pirate editions of it in 1822. Lawrence took legal action to protect his copyright. The Lord Chancellor at the time, Lord Eldon, refused to extend legal protection to Lawrence's rights as an author, because the book in question was *blasphemous*.

Brown did not, it seems, have a radical cell in his body. He had already experienced the effect that his father's loyalty to dissenting opinion

---

* For an excellent account of this debate and Lawrence's biography, see chapter 1, The Vitality Debate, 1814-1819, in *Shelley and Vitality* by Sharon Rushton, (2005), Palgrave MacMillan, London.

had on life. (See above, chapter 1.) He would have been most uneasy at being tarred with the brush of sedition. He and the gentlemen-scientists of his social circles surely would have been appalled that his discovery could be thought to give support to radicals and revolutionaries. For all his great scientific gifts, Brown was not the sort who would have been able to or would have wanted to stand up to conservative authority. He would surely also have felt most unhappy at being out of step with his scientific colleagues. During this period campaigners were going to prison, being transported to Australia, and even being executed for treason. The political atmosphere must have been enough to terrify anyone, let alone a quiet soul like Brown. Doubtless it was meant to terrify. Clearly Brown, not being a revolutionary, was not a target, but the enormous pressure to conform and the imagined or real threat from the poor and disenfranchised could have been used against him, and, I argue, probably was. He just could not be seen to be lending support to such ruffians as the radicals and the mob.

A strong connection between this social history, Brown's own personality, and the interpretation of his experiments with pollen and other materials and his discovery of the active molecules begins to emerge.

I can imagine readers raising their eyebrows at the next step in my speculative argument, but I shall push on in the hope that at least a few readers have enough tolerance and curiosity to follow me to the end.

Spontaneous generation is associated with godless materialism and revolutionary opinions and actions. Brown's initial interpretation of his findings is that he has observed this very process or at least the first steps of it, life organising itself into existence from below. (Even now in 2012 this is an idea that profoundly disturbs many people, especially scientists.) And the same people who find this so abhorrent in Brown's day are also afraid of rebellion from below by the undisciplined, spontaneous masses, the toiling, dirty millions, so precariously controlled by repressive government.

If sceptical contemporary readers sneer at the parallel, the comfortable, respectable classes of Brown's day certainly saw it as obvious, especially Christian, conservative scientists. Unconsciously, perhaps even consciously, they saw the common functioning principle that Reich observed between spontaneous generation, (his discovery of the bions and bionous disintegration), and the spontaneous, involuntary urge towards self-regulation and democracy by the masses. If there is such a spontaneous urge towards democracy and self-government, then the advocates of an authoritarian control of society and gross privilege, such as the governing classes of Brown's day, are faced with a huge problem, confronted with the possibility of endless division and conflict. Either the rebelling masses get their way and establish democracy, or thoroughgoing, out-and-out suppression and control from above is imposed.

As my orgonomic colleague, Philip Bennett points out,[19] within three years of each other, Reich wrote two papers on these topics with titles that start with exactly the same words; *The Natural Organisation of Work in Work Democracy*[20] and *The Natural Organisation of Protozoa from Orgone Energy Vesicles*.[21] He, too, clearly saw the connection, even if at this stage of his work he had not clearly formulated the *common functioning principle.*[22] *

Given this background in 1828, it is not too fanciful to claim that some influential gentleman-scientist, possibly someone who was not even a scientist, someone interested and socially aware, must have come up to Brown one quiet evening in the late Sir Joseph's library and pointed out how damaging it would be to his reputation and encouraging to the radicals and the mob to be seen to be advocating such a disreputable, godless doctrine as that of spontaneous generation. To do so would be to give ammunition to one's enemies. Given Brown's diffident character, this persuasion would not have needed to be even a slap on the wrist. A hand on the elbow, a few quiet words in his ear, would have been enough. And so we have the strange document, the *Additional Remarks*.

If you think this is stretching things far too far and wish to ignore this chapter of conjecture, you must still account for Brown's description of pulsatory, vermiform motion and the agglomeration of his molecules into motile bodies.

---

* In orgonomy the common functioning principle (CFP) links two processes in nature where the energy functions, eg. pulsation, expansion, contraction, discharge, stasis, movement, superimposition, are the same. It therefore often links apparently disparate processes that orthodox science sees as having nothing in common at all. See EGD for more details.

[1] Mabberley D J (2004); entry on Brown in *Dictionary of National Biography*, volume 8, page 110, Oxford University Press, Oxford.

[2] Darwin C (1881); Autobiographies, page 60, re-published (2002) by Penguin Classics, Penguin Books, London.

[3] Mabberley D J (1985); Jupiter Botanicus page 159. Cramer, Braunschweig, and British Museum (Natural History), London.

[4] Mabberley D J (2004); op cit, page 110.

[5] Ford B J (2009); Charles Darwin and Rober Brown – their microscopes and the microscopic image, *Infocus*, Issue 15, September, 2009, Oxford. There is also a film of Ford's project available via Ford's website – www.brianjford.com and on YouTube. Type 'Brian J Ford' into the search window.

[6] Strick J (1992, unpublished); Robert Brown and Brownian Movement: Radicalism, Spontaneous Generation and Microscopy in Nineteenth Century England, pages 1-2.

[7] Needham J T (1748); A Summary of some late Observations upon the Generation, Composition, and Decomposition of Animal and Vegetable Substances, *Phil Trans. Roy. Soc.*, volume 45, page 665, London

[8] Hilton B (2006); A Mad, Bad, and Dangerous People? New Oxford History of England: England 1783-1846, page 398,Oxford University Press, Oxford.

. [9] ibid; pages 586-587.

[10] Figures worked out approximately myself from statistics given in Gardiner J and Wenborn N (eds) (1995); History Today Companion to British History, Reform Act 1832, page 639 and Population, pages 610-611, Colin and Brown, London. .

[11] ibid, page 184.

[12] Hilton B (2006); op cit, pages 580-581.

[13] ibid, page 629.

[14] ibid, page 355

[15] Farley J (1977); The Spontaneous Generation Controversy from Déscartes to Oparin, page 43, Johns Hopkins University Press, Baltimore.

[16] Desmond A (1994); Huxley: The Devil's Disciple, page 161, Michael Joseph, London.

[17] Mabberley D J; op cit, page 265-266.

[18] Lawrence W (1819); Lectures on Physiology, Zoology, and the Natural History of Man, delivered at the Royal College of Surgeons, J Callow, London.

[19] Bennett P (2010); Personal communication.

[20] Reich W (1939); Die Natürliche Organisation der Arbeit in der Arbeitsdemokratie, Politisch-Psychologosche Schriftenreihe, No 4, (The Natural Organisation of Work in Work Democracy, Political-Psychological Writings, No 4), Sexpol-Verlag, Oslo.

[21] Reich W (1942); The Natural Organisation of Protozoa from Orgone Energy Vesicles, *International Journal of Sex-Economy*, I, 1942, Orgone Institute Press, New York, included in **CB**, chapter II, 6.

[22] EGD**;** discussion of the common functioning principle, pages 103-106. See also CS, passim.

Chapter 5 Motivations? The Assumptions of Mechanistic Science

You may be wondering why these stories are repeated so endlessly and so unquestioningly and why students swallow them so readily. Is there a motive? If science is really science, why is one explanation so much more acceptable than another? Alas, that is the point. Science is not the noble, altruistic, selfless search for truth that it presents itself as. If it were, there would be no controversy. One experiment would be as good, as interesting, as relevant, as acceptable as another and the only thing that mattered would be the outcome and the findings, the conclusions to be drawn from them. But this is not the case.

Lay-people almost certainly do not know this, but those working on the fringes of science and a few within its mainstream know very well that a research project or a theory that suggests the existence or a possible effect of a life force will not be received in the same way as a research project or theory that specifically excludes such a possibility. Any interest whatsoever in the possibility of the existence of a life-energy is branded as flat-earth mysticism by conventional scientists. Mysticism is, in their book, the worst crime in the universe. A good example of the response of conventional scientists to the first sort of research is that of the journal *Nature* to the work of the French scientist Jacques Benveniste. A deputation from the journal went over to his Paris laboratory to check his work. The deputation included a magician: there was clearly a suspicion that his findings were deliberately fraudulent.[1]

Benveniste carried out a research project that seemed to demonstrate that water molecules have some capacity for memory. Water from a stronger medical solution was diluted and diluted in standard homoeopathic fashion. According to the standards of ordinary science, this water could not possibly have any medicinal effects, because it no longer contained any actual molecules of the medicinal ingredient. According to Benveniste's findings it appeared to still have some medicinal effect. He apparently was at the time beginning to develop a hypothesis concerning a possible memory function of water. Such an idea enrages conventional scientists, for reasons best known to them. It is, after all, only a scientific hypothesis, which can be tested and confirmed or discarded as not true.

This brings us to one of Reich's discoveries that seems to be outside the scope of this book and my main argument, the discovery of muscular armouring. However, it is quite unreal to divide Reich's work on the discovery of the life energy into compartments and to separate one part from another. They are all connected by a red thread, the existence of the cosmic orgone energy and its functions and the consequences of our culture's denial of its existence, both socially and individually.

This denial takes place in each individual in our society as he or she, as a small baby, has to deny in him/herself the pain of their unsatisfied primary needs. We deny this pain by preventing the free, natural, spontaneous movement of our own orgone energy. This free movement of our orgone energy is experienced as needs, sensation, and emotion. We prevent the perception of these needs by contracting physiologically and forming involuntary muscular tensions, what Reich named *muscular armouring.*[2] A basic law of muscular armouring is that once the free movement of energy has been inhibited by it, we automatically feel anxiety, if not outright fear, when the orgone energy tries to move again and flows back to enliven the hitherto dead areas. We see clients in orgone therapy reliving this fear as their armouring starts to give way. Before the armouring gives way, the client will feel disturbed, threatened, and insecure, as these strange (to the client) feelings try to push through. I can only assume that scientists are experiencing a similar effect, when they have such strong reactions to the innocuous suggestion that there may be a life-energy and that it may be possible to test its effects experimentally.

Though my explanation is speculation, it is an undeniable fact that orthodox biologists and indeed almost all scientists feel extremely uncomfortable with the idea of a life force active in nature. They invariably become angry when one suggests that there is such a force. People become angry in this way only when one is pressing an emotional button. An unbiased reaction would be to consider the theory or experiment on its merits. James DeMeo, the US orgonomist, reports a physicist shouting out, 'That can't happen!', when he demonstrated the lumination of orgone energy in a high-vacuum tube that had been charged in an orgone accumulator for a long time prior to the demonstration.[3] C O R E has a similar vacuum tube that lives in a small orgone accumulator within a larger one and the tube luminates visibly in the dark more or less to order when stroked. Many different people have seen it luminating in this way. This is one of the simplest and most direct demonstrations of an orgone energy effect.[4] If the tube has not been charging up inside an accumulator for a period of weeks it does not luminate.

There are plenty of other simple demonstrations of orgone energy effects that the novice orgonomist can reproduce far more easily than those under discussion here, the bion experiments of Brown and Reich. Two easily reproducible ones are the temperature rise in a human who sits in a large accumulator for a while in good atmospheric conditions* and the seed-germination experiment. Both of these can be repeated by a child of ten or so without difficulty. All these simple experiments or demonstrations seem to threaten conventional scientists and they frantically look round for ways of explaining them away or ridiculing them.[5]

I leave these thoughts and suggestions on the table for the curious reader who will doubtless continue the debate within their own mind, if not in public. This is an important debate that is most of the time pushed under the carpet. I hope the effect of this book will be to pull it out into the daylight, though I am not too hopeful. The capacity of science to keep the lid on unwelcome discoveries and facts is prodigious, and doubtless it will find a way of neutralising the information within this book. It does not need to do much. All scientists need to do is to keep quiet and to wait until the fifty people who buy and read this book have passed on or have just forgotten what they have read. Another way of dismissing the findings contained in this book will be to discredit me, the author, by publishing the fact that I have no scientific qualifications. I have no wish to conceal this fact and present it to my opponents as free ammunition. The wider public are almost as reluctant as orthodox scientists to face these, to them, unwelcome facts.

The Assumptions of Mechanistic Science

While writing this story, it dawns on me that many readers, especially the large numbers in the middle, neither rigid mechanists nor gullible new-agers, will wonder why my theory should be at all controversial. He's only a keen amateur with a bee in his bonnet, isn't he, after all? He's only trying to find out exactly what Brown did and wrote. Yes, that's all true and I have committed no crimes, but… wait until the book comes out! By readers in the middle I mean those on an imaginary spectrum of varying responses to this book. At one end I imagine a tiny number of scientists or scientifically trained readers who want to throw it in

---

* 'Good' here means high pressure atmospheric pressure and low humidity. As humidity is always relatively high in this country, orgone devices need a much longer charging period and tend to be less markedly effective here than they are in drier environments.

the rubbish-bin and some of whom probably actually will. At the other end I imagine a small number of those at the other cultural extremity, readers of *Nexus* and *Fortean Times* and so on, who jump ecstatically at anything that seems to be critical of conventional science and acclaim it unquestioningly. And in the middle come the large majority of readers with no powerful prejudices, who are just interested in the real history of two important scientists and the discoveries they made.

It says so much for the total triumph of modern science that the word is never qualified now: science is simply science, and to most people there is only one science, what passes for science now. Most readers will, therefore, be mystified by the word mechanistic here.

Science is indeed now so completely mechanistic, mechanistic and proud of it, in its basic principles and assumptions, that it does not need to say so. This has not always been the case. At certain times in history very different types of science based on very different basic assumptions have been dominant. For example, in the last half of the eighteenth century western medicine and biology were deeply vitalistic in orientation, particularly in Germany and France. Vitalism was an active influence well into the nineteenth century in Germany and Britain.[6] Mechanistic science, in particular medicine and biology, came to the fore in the nineteenth century and it has since ousted virtually all other models of science. The belief that mechanistic science is the only science, is what science is, is now so deep and unconscious that most modern histories and reference works do not even contain entries for either mechanism or vitalism in their indexes.[7] I have been able to find a definition only in a non-scientific dictionary.[8]

Mechanistic science assumes that all life, the universe, and the inanimate world, can be explained by the general principles and forces of physics and chemistry. Matter can be neither created not destroyed; energy flows from a higher level to a lower; in biology, evolution occurs by the impersonal, indifferent process of natural selection, in which nature selects beneficial random variations that allow individuals and species to survive better than their competitors; creatures, even organs, do not develop with a purpose in view; changes are random and beneficial ones survive, harmful or useless ones are automatically culled by natural selection and do not survive; there is no purposeful organising force in nature; there is no difference between chemicals made within biological organisms and those made outside of them; there has been no creation of new life. Life has simply evolved according to the natural principles of physics and bio-chemistry and natural selection.[9]

Although it is not at all inevitable to draw the next conclusion from the principles of mechanistic biology, it is in fact taken as read by all present-day scientists - the conclusion that life is not now starting from scratch, that life in some now completely unknown way started off once as the result of a random accident or accidents in conditions that were so different from those pertaining today that it cannot now happen again.[10] It is a heresy to claim that life is still starting itself off now. Such an occurrence is known as spontaneous generation and is now ruled out of court by science without further debate. The occurrence of spontaneous generation is always considered by historians of science to have been proved impossible by the experiments of Louis Pasteur in 1859. These employed the classic swan-necked flasks and their contents of sterile fluids that we see pictures of in almost all undergraduate textbooks of biology and microbiology.[11] However rational and calm one is about this question, when any researcher or theorist claims that spontaneous generation might occur now, conventional scientists reach for their machine guns and hand-grenades. In 1981 the then editor of *Nature,* John Maddox, described Rupert Sheldrake's book on morphogenesis as 'the best candiate for burning there has been for years.'[12] The reaction is irrational and aggressive. So, I can only conclude that this possibility means a lot more to ordinary scientists than it does as a simple scientific proposition. (What are they afraid of?)[13]

Above all there is definitely no such thing in nature as a vital force, call it what you will. Anyone who dabbles with even a possibility that there may be such a force is automatically rejected as unscientific. They will be branded as mystical, the worst crime in mechanistic science. However meticulous a scientist's experimentation is, however carefully his results have been written up, the very connection with a life force is enough to have him axed, expelled, and dumped by the community.

So… although the arguments of this book rest on experimentation, this makes no difference. By arguing that there is a life force, orgone energy, and that, in certain commonly occurring circumstances, it can organise dead matter into transitional forms, linking dead matter and the living, I know I put myself beyond the pale. My arguments will not be treated fairly. Few, if any, scientists within the conventional scientific community will take the trouble to repeat these experiments and make up their minds for themselves. That is why I have included, I hope, enough information to enable interested and motivated amateurs to repeat the experiments themselves. They are the only readers who will be open-minded enough to give these experiments a fair trial. It is strange but true that those most suited to put these findings to the test, scientists with training in microscopy and microbiology, are the least likely to carry out these tests.

The Implications and Consequences of these Discoveries

So…where does it leave us, if my claims are confirmed? What are the scientific implications of the existence of a life-force that brings bions into existence? What are the conclusions to be drawn? I could simply present my findings and leave readers to draw their own conclusions. Doubtless, many will already have come to some realisations concerning science after reading so far. I will spell out the conclusions that have occurred to me while I have been involved in this work. If you are a naïve believer in the goodness of science you may be surprised to read what comes next. If you are sceptical towards science you will not be the least bit surprised by what I say in this chapter.

The first thing that occurs to me is that, if it is true that life is all the time organising itself into existence, the basic assumptions of Darwinism are wrong, plainly and simply wrong. This truth, if confirmed, is enough to set the Darwinists reaching for their machine guns. The philosophical assumption behind Darwinism is that life is a completely random event and that evolution is driven by random events, accidental mutations. The idea that there is a purpose and direction to evolution and that life is 'trying' to do something, is trying to move in one direction rather than another, is anathema to Darwinists. In the jargon of debates in biology, it is branded as *teleological*. That means that something has an end purpose and its activities are directed towards the realisation of that purpose. It is a basic assumption of Darwinism that life and all change in living forms are entirely accidental.[14] But the discovery that life is developing involuntarily all the time suggests that there is a direction, a tendency towards life. If there is, then surely the same urge towards life must be at work within organisms, once they exist. I think we can make a good case for that theory, though I don't want to go into the details of that argument here. I have always wanted to keep this book simple and within the realm of the ordinary and the down-to-earth.

An unspoken assumption of many people who 'believe' in a life-force and the primary importance of life and nature, is that there is somewhere, somehow, a natural order of things and we interfere with that natural order at our peril and at its peril.[15] Again, such an assumption is anathema to science and most scientists. The denial that such an order exists allows scientists to interfere with the genetic structure of organisms and the natural cycles of the planet without a qualm. If there is no natural order, we can do what we want, can't we? When lay-people accuse scientists of 'playing God', that is exactly what they are referring to in an instinctive way. These attitudes to life are often intuitive and inarticulate, even amongst scientists, but particularly amongst those who acknowledge a natural order of things and a life-force.

Experiments that demonstrate clearly and unarguably that there is indeed a life-force and therefore a natural order of things threaten the rule of science at a very deep level. Scientists know this instinctively and recognise any model, theory or, even worse, experiment, that threatens to confirm this. This explains why Brown's tentatively vitalistic explanation of his findings was noticed and squashed by someone and why Reich was so bitterly persecuted in his lifetime, and even beyond the grave. Science's intuitive awareness of what it calls vitalism and mysticism is, like a smoke-alarm, sensitive down to one part per million. Certain questions set scientists' alarms off even before the question is finished.

The question never asked is - why it is so terrible to argue for the existence of a life-force? What would be wrong if one were to demonstrate that there is in fact such a force? It would only be a scientific fact. I am sure that if we put this question to a scientist publicly, he would squirm and wriggle and show acute discomfort, but would not be able to give a rational argument for his position. I can imagine such a scientist running from the room screaming, especially if he were confronted with incontrovertible experimental evidence. We are here bumping up against the scientist's armouring, his defences against his ability to feel the movement of his own life-energy within himself, the great denial that is part and parcel of western culture, in particular western scientific culture.

---

[1] http://www.buzzle.com/editorials/3-29-2005-4752.asp

[2] FO; chapter VIII, 1, Muscular Attitude and Body Expression..

[3] DeMeo J (2002); *Pulse of the Planet*, No 5, page 255, Ashland, Oregon.

[4] CS, photograph of an orgone-charged vacuum tube luminating. Readers following up this reference should note that the photograph is only reproduced in the hardback edition of this publication.

[5] Gardner M (1957); Fads and Fallacies in the Name of Science, Dover, New York.

[6] Fuchs T (2001); The Mechanization of the Heart – Harvey and Déscartes, Vitalism and Mechanism between 1700 and 1850, University of Rochester Press, Rochester, NY.

Reill P H (2005); Vitalizing Nature in the Enlightenment, University of California Press, Berkeley, California.

[7] Gribbin J (2002); Science A History 1543-2001, Penguin, London.

[8] Pearsall J, Trumble W (eds) (2002); Oxford English Reference Dictionary, *mechanism*, page 897, Oxford University Press, London.

[9] De Duve C (1995); Vital Dust: The Origin and Evolution of Life on Earth,

[10] Purves W K et al (1995); Chapter 24, the Origin of Life on Earth, pages 518-519, Life: The Science of Biology, Sinauer Associates, Sunderland, Massachusetts.

[11] Madigan M T et al (1997); Brock Biology of Microorganisms, page 22, Prentice-Hall, Upper Saddle River, New Jersey.

[12] Maddox J(1981), cited on the cover of Sheldrake R (1983): A New Science of Life, Granada, London.

[13] Milton R (1994); Forbidden Science, chapter 5, Animal Magnetism.

[14] Darwin C (1859); The Origin of Species, reprinted in Penguin Classics (1985), Penguin, London.

[15] Balfour E (1943); The Living Soil, passim, Faber and Faber, London, reprinted (2006); The Soil Association, Bristol.

Chapter 6 The Drip, Drip, Drip of Deception

It seems inconceivable that an investigation such as Brown's, well documented and easily replicated, can end up being taken to demonstrate the opposite of what it actually demonstrates. That is, however, the outcome of almost two hundred years of history, almost two hundred years of misquoting, distorted handings down, and the unquestioning repetition of other people's neutered versions of the original.

How has this come to be? I cannot trace this process, step by step. It would produce a massive and doubtless boring tome. Here I aim simply to show the sort of things that scientists have been doing to get this distortion going and, once the myth had been started, to keep it going. It now has a life of its own and just grows and grows, modern utterances barely having any connection at all with Brown's original text, as we shall see. I have come across all the writers mentioned here by accident. I was not deliberately following a trail of evidence. I doubtless could have chosen a completely different group of scientists or works that over the years have also contributed to the myth. Those mentioned here are just the ones I have tripped over in my travels. This project started with Einstein and his *The Evolution of Physics.*[1] I had never even heard of this book and happened upon a cheap copy in the Preston Oxfam bookshop. It was that book which opened my eyes to the blatant deception that was being perpetrated by Einstein, and presumably by others. But I shall deal with the evidence chronologically, so we shall come to him later.

Obviously the whole process was given a hefty push from behind by Brown's own nervousness and his retraction contained in the *Additional Remarks*. This must have left the question of exactly what Brown had observed quite undecided. If he had not been claiming that his molecules were alive or a transition between non-life and life, what was he claiming? His immediate mention of De Buffon and Needham surely demonstrates that he had decided that he had unintentionally confirmed De Buffon's theory[2] that there was a basic element to life, which he, De Buffon, called an organic molecule, and Needham's claim that he had obtained living forms from sterile preparations.[3] These two worked together and were unambiguously associated with a belief in spontaneous generation. To mention their names, coupled, where Brown does mention them, surely leaves little doubt as to his initial interpretation of his findings. His *Additional Remarks* can surely be nothing but a change of heart, if not a downright retraction. His ambivalent attitude gave other people the right to decide, if they wanted to, what they thought Brown was claiming. Possibly this explains the limbo in which his discovery floated until it was taken over and interpreted by physicists and mathematicians.

I was already quite familiar with Brown's original text when I came across Einstein's *The Evolution of Physics*. I had not heard of this book and was not searching for it. (I spend a good deal of time in secondhand bookshops and on line searching for rare books that are important to orgonomy for C O R E's library.) I knew that Einstein had worked out the mathematics of 'Brownian motion' and that his calculations were taken in science as the final word. Would this little book have anything about it? There was an entry in the index, pages 63-67. I looked them up there and then in the bookshop.

I found a quotation taken from Brown's article with the commonly seen dots at the end of a sentence, meaning that the writer is leaving something out of the original text. This is a common practice, something I have done myself, and in itself there is nothing sinister about it. It all depends what you are leaving out. I knew the original text well enough to suspect that the lines that Einstein or his assistant had left out were those about the pulsatory motions that Brown describes. I bought the book, so I could take it home and check up on my hunch. I had an intuition that Einstein and Infeld were deliberately pulling the wool over readers' eyes. And so they were. Here is the quotation as it stands in the book:

> While examining the form of these particles immersed in water, I observed many of them evidently in motion … These motions were such as to satisfy me, after frequently repeated observation, that they arose neither from current in the fluid nor from its gradual evaporation, but belonged to the particle itself.[4]

So what have they omitted? What lies behind those dots?

> …their motion consisting not only of a change of place in the fluid, manifested by alterations in their relative positions, but also not infrequently of a change of form in the particle itself; a contraction or curvature taking place repeatedly about the middle of one side, accompanied by a corresponding swelling or convexity on the opposite side of the particle. In a few instances the particle was seen to turn on its longer axis.[5]

This is a clear description of some sort of pulsatory motion, expansion and contraction. The words could not be clearer. They are an embarassment to the whole 'Brownian motion theory' as espoused by modern science and so they are ignored, omitted. But not only by Einstein. His example is probably the worst, the most blatant, but he has got plenty of company. To be fair to Einstein, Infeld was probably responsible for the deliberate omission and Einstein did not even know that he had made it. But if your name is down as the main author of a book, surely it is up to you to take responsibility for the contents and to check on things like that? Information on the writing of *The Evolution of Physics* suggests that Infeld was in fact the main author and that Einstein was doing him a favour by allowing his name to be used. According to Isaacson, a recent biographer of Einstein, the writing of the book was Infeld's proposal and Einstein happily accepted it. Infeld was finding it difficult to obtain a post in America and Einstein was keen to help him.[6]

Needless to say, there is no mention whatsoever of another process that Brown describes and which Reich also observed, the agglomeration of the active molecules or bions into larger forms which also contract and move spontaneously in a way which cannot conceivably be 'Brownian motion.' Here is Brown's description:

> In many of the substances examined, especially those of a fibrous structure, as asbestos, actinolite, tremolite, zeolite, and even steatite, along with the spherical molecules, other corpuscles were found, like short fibres somewhat moniliform, whose transverse diameter appeared not to exceed that of the molecule, of which they seemed to be primary combinations. These fibrils, when of such length as to be probably composed of not more than four or five molecules, and still more evidently when formed of two or three only, were generally in motion, at least as vivid as that of the simple molecule itself; and which from the fibril often changing its position in the fluid, and from its occasional bending, might be said to be somewhat vermicular.

There we have it, the use of the word *vermicular*. I tracked down Brown's original 'slip' on which he wrote his notes on these investigations[7] and he uses the same word there, so it is obviously not an afterthought. Perhaps many modern readers don't even recognise this word and pass it by. It means worm-like and is a further clear reference to pulsatory motion, as recognised, described, and illustrated by Reich later in *Die Bione*. Further:

> In some of the vegetable bodies burned in this manner, in addition to the simple molecules, primary combinations of these were observed, consisting of fibrils having transverse contractions, corresponding in number, as I conjectured, with that of the molecules composing them; and those fibrils, when not consisting of a greater number than four or five molecules, exhibited motion resembling in kind and vivacity that of the mineral fibrils already described, while longer fibrils of the same apparent diameter were at rest.

It is utterly impossible to account for these movements by 'Brownian motion.'

The next study of Brown's experiments that I came across is an academic article dating from 1971, *The Discovery of the Brownian Motion*, by Peter W van der Pas.[8] Van der Pas has no doubts at all and strides off, describing Brown's discovery as definitive but unexplained.[9] He actually mentions Brown's awkward references to the agglomerations that he observed as 'built up from spheres', but gets round the problem by simply stating that Brown had not actually seen these items. He was mistaken because he was working with 'an imperfect lens at the borderline of its magnifying power.'[10] This is a completely illegitimate ploy that is often used by critics to dismiss other workers' inconvenient findings. Notice, van der Pas does not say that he has repeated Brown's experiments using his, Brown's, own microscope or a comparable instrument and found that… He simply states an assumption that he must have been… The same assumptions were thrown at Reich. His bions 'must be' Brownian motion or contamination.[11] The accusation of 'Brownian motion' was also thrown at H C Bastian by T H Huxley in the same way in 1870.[12]

This 'he must have been….' is an old tactic and was well established before Brown wrote his 1828 paper. As I revise this chapter in early 2011, I have come across yet another example of this method of dismissal. It dates from the later eighteenth century (1787), from before Brown's paper. But, yet again, it is an attack on the protagonists of spontaneous generation. George Adams the younger (1750-1795), well-known in his day as a maker of microscopes, rubbishes the work of De Buffon and Neeedham in this way, alleging that they had probably never seen the items they were describing and were confusing animal and vegetable remains with spermatozoa.[13] Much more recent scholarship shows that in fact De Buffon had described all too well what he had seen and that his images were clear and undistorted. A US researcher, Philip Sloan, has checked DeBuffon's findings with his, De Buffon's, own instrument and confirmed them.[14]

Brian J Ford has since demonstrated that Brown had indeed really seen what he was describing and that his images were excellent.[15] It is common to dismiss claims of the early microscopists working with simple single-lens instruments with the assumption that the optics of their instruments were very inferior and that the images obtained must therefore have been very poor. Ford disproves this conclusively as we shall see below.

The rest of van der Pas's paper is devoted to the technicalities of the discovery and how in fact another observer, Jan Ingenhousz (1730-1799), had described 'Brownian motion' in passing in a paper published in 1789 on the use of the microscope and improvised coverslips to prevent the evaporation of fluid from a slide.[16] (Commercially manufactured coverslips were not in common use until about 1840, as far as I can ascertain.) Ingenhousz's paper is attached to van der Pas' paper and is a useful item for any student who wants to know the whole story of 'Brownian motion.'

We would expect to find an accurate version of events in a thoroughly researched, academic biography of Brown written by a professor of botanical sciences. There is such a biography of Brown available in English, *Jupiter Botanicus*, by Professor David Mabberley.[17] It is in fact the only biography of Brown to have been written. Professor Mabberley is now Keeper of the Herbarium, Library, Art, and Archives at the Royal Botanic gardens, Kew. He has held numerous academic posts over the years and is obviously a distinguished scientist. We can expect him to know his subject. He starts his 'author's note' by arguing that Brown should be much better known than he is today for his great achievements and discoveries in the field of botany. He is surprised that there is so little written about Brown, given the importance of his botanical discoveries. That he was the first microscopist to describe the bions and bionous motility makes his achievements even greater. I completely agree with professor Mabberley that Brown's reputation is strangely limited in comparison to those of, for example, Newton, Darwin, or Gallileo, who are all household names, even amongst people who have little knowledge of science. For some reason botany does not possess the glamour of astronomy, mathematics, and evolution.

So what does professor Mabberley have to say about Brown's description of pulsatory motion amongst some of his molecules and the motile agglomerations that he observed and the description of which I have quoted above?

Nothing. They are not mentioned.

We now come to the work of Dr Brian J Ford. I hesitate to cross swords with such a distinguished practical microscopist and to risk making an enemy of him. Ford has done more than anyone else to dispel the myths about Brown's work, the simple microscope, and what it can achieve.[18] His writings on the subject are clear and interesting and indispensible reading for any serious student of the history of microscopy. He continues to throw light on the use of the simple microscope[19] and doubtless we can expect more informed articles by him on the topic.

But he, too, probably more familiar with Brown's research than any other living scientist, ignores, as far as I can see, all reference to pulsatory motion and to how Brown's active molecules form motile agglomerations. Ford is such a prolific writer that I have certainly not read everything he has written on this topic. The most definitive article by him on Brown's work appears to be *Brownian Movement in Clarkia Pollen: a Reprise of the First Observations*.[20] In this project Ford repeats Brown's investigations using Brown's own simple microscope, that is exactly the same instrument as they were originally made with. His intention was to lay the ghost, once and for all, of the critics who disbelieved Brown's report on the basis of ill-founded assumptions concerning the abilities of the simple microscope. As part of this investigation he made a film of 'Brownian motion' through the lens of this instrument.

An inconsistency that I notice in this report is that Ford, quite rightly, dwells on the fact that any experienced microscope observer will readily recognise Brownian motion and that it has various umistakable qualities. Now if he has repeated Brown's pollen investigation exactly as described by Brown, he must surely have noticed other motion amongst the smaller particles, bionous motility and the motile bionous agglomerations, which we inevitably find when pollen ruptures in water. If he had checked further, as did Brown, examining the behaviour of finely ground minerals in water, for example, fossilised wood and ground stone, he would have noticed this too. But about these forms and their motion we find not a word.

Now we come to the more lightweight misinformation, items that are almost on a par with gossip, things that people apparently have just repeated, without checking that they are true. The first of these I found when out of curiosity I looked up *Brownian motion* on BBC – h2g2. The article[21] gave a conventional version of events but stated that Brown had been using a water immersion lens. As Ford demonstrates, this is definitely untrue and the statement shows how people repeat loosely researched versions of events, perhaps versions not researched at all. (Amici's water-immersion objective was not developed until 1853 and would, anyway, have been completely irrelevant to the simple microscope used by Brown.[22]) Yet again in this article there is no reference to the pulsatory motion or the motile agglomerations described by Brown. But we expect that by now.

There is plenty of information about Brownian motion on YouTube and I have followed up some of these contributions. That led to an item on Brownian motion which was 'recommended for you' from Sixty Symbols, a collection of videos on various scientific topics put together by a professional documentary film-maker, Brady Haran, with the collaboration of academics from Nottingham University. This video featured Professor Roger Bowley, professor of physics in the Department of Physics and Astronomy. To be fair to Professor Bowley, he is not a biologist and so can be forgiven for his rather bowdlerised version of Brown's experiments. His speciality is statistical mechanics and he has written the best-selling ever student textbook apparently. (I have read that somewhere, but have been unable to check it and can unfortunately give no reference for the claim.) Even so, it seems strange that a specialist in his field has apparently not read Brown's original paper.

According to Professor Bowley's version of events,[23] Brown wanted to know whether pollen grains were alive and sprinkled them onto water. To check his findings he did the same with sand and other dust particles. These remarks are almost irrelevant to the meat of his presentation on Brownian motion, but they do show how people, even academics, who should know better, throw around versions of events without checking facts. Imagine the picture that a student will get from this who has hardly heard of Brown and certainly not read his original paper.

Continuing our list in order of publication, we come to quite a minor example, the reference to Brownian motion in Dorling Kindersley's *Science*, one of their giant pictorial historical reference works. (This is a beautiful, fascinating publication and I own a copy, which I refer to frequently.) In the 'time line' for the years 1700-1890 the entry for 1827 states: Robert Brown defines Brownian motion.[24] Now any honest reading of Brown's *Active Molecules* paper makes it quite clear that the paper does not define anything at all. Brown describes and puts forward a tentative explanation. It is quite clear that Brown is unsure of what he is seeing and leaves the matter open. If he defines anything, he leans towards a definition of his molecules as a basic element in all living matter.

> Reflecting on all the facts with which I had now become acquainted, I was disposed to believe that the minute spherical particles or Molecules of apparently uniform size…were in reality the supposed constituent or elementary Molecules of organic bodies, first so considered by Buffon and Needham, then by Wrisberg with greater precision, soon after and still more particularly by Müller, and, very recently, by Dr Milne Edwards, who has revived the doctrine and supported it with much interesting detail. I now therefore expected to find these molecules in all organic bodies; and accordingly on examining the various animal and vegetable tissues, whether living or dead, they were always found to exist; and merely by bruising these substances in water, I never failed to ascertain their apparent identity in size, form, and motion, with the smaller particles of the grains of pollen.

My final items come from two children's books, the Usborne *Illustrated Dictionary of Science*[25] and *Das große Kosmos-Buch der Mikroskopie*.[26] The brief account in the first book is little better than gossip. Within a *Kinetic theory* box under a heading *Brownian motion* there is a picture of a modern monocular microscope. A circle shows a blown-up view of what the observer is seeing on the slide and inside this circle are drawn pollen sacs with criss-crossing arrows between them, implying the random, oscillatinging Brownian motion of these sacs. The writer calls them 'grains' but the drawing is clearly of sacs and pollen sacs are far too large to show Brownian motion. We can see pollen sacs as dust and separate them visually with a 10x magnifying glass. Clearly the writer of this item has not checked up on any facts at all.

The account by Kremer in the Kosmos book is presented more scientifically, but there is, yet again, no reference to Brown's observations of pulsatory motion and agglomerations of his active molecules. Kremer claims that Brown's observations confused contemporaries by leading them to think he had discovered the *monads* thought to exist by Plato and Leibnitz. In fact Brown himself suggested this, not his contemporaries, though he mentions neither Plato or Leibnitz. He does mention De Buffon and Needham, both protagonists of spontaneous generation who thought they had seen the basic elements from which all life was composed. It sounds as if Kremer has got his information from other people's versions of Brown's paper. Kremer makes matters worse by advising his trusting young readers to use wheat-flour to demonstrate Brownian motion for themselves.[27] If we put a tiny pinch of wholewheat flour on a slide, add a drop or two of water, cover the mixture with a coverslip, and examine it at 800x, we see…bions!

This version of history continues in all science writing. Reading these summaries, which occur all over the place in science books, one would never suspect how tentative Brown's voice is. No, this is 'The Discovery of Brownian Motion.'

From reading these versions of Brown's investigations, no student of microscopy or biology would suspect that a hugely important part of his findings is completely ignored by everyone who has written about his discoveries since his death. In the many booklets on orgonomy that I have written for C O R E over the years I advise readers not to trust my second-hand versions of Reich's work and to go back to the originals and to read and re-read them. The same goes for Brown. I do not suppose one student of biology or microbiology in a thousand, even ten thousand, ever thinks of reading the actual text, as they listen to some professor delivering the conventional, neutered version of his paper. These professors should be urging their students to both read the original text and to repeat the experiments described in it. If I ever run an orgonomic microscopy course, I will set students the task of repeating experiments by the early pioneers, such as Leeuwenhoek, Spallanzani, or Needham, not to mention Brown, of course. I have done this myself and learnt a great deal while doing so.

Can we say anything to excuse these omissions? Can we explain them at all, except as a blatant attempt to deceive readers? I have tried to put myself in the minds of people who can read a paper such as Brown's and completely overlook things described in it that are so crucial to his observations. It is as if someone were to walk out into a garden and cheerfully stride across a beautifully sown flower-bed, bursting out through a hedge without apparently even noticing the damage they have done. Can't you see…? Haven't you noticed that…? the simple, flower-loving soul, the creator of the bed, gasps, as their beautiful leaves and stems crunch under the philistine's boots. But no, he is off and down the lane before onlookers can open their mouths. Someone else follows. Soon the act of vandalism has been repeated so often that eventually there is a trampled pathway across the border and through the hedge. After a while no-one even notices any longer that there is not really a path there and never has been.

Working in orgonomy, one gets used to seeing things that most people never notice. It may be that only after a goodly immersion in Reich's bion experiments does one notice the descriptions of pulsatory movement that we find in Brown's paper. Having seen forms and agglomerations exactly as described by Brown, one is primed to notice his descriptions as familiar and immediately recognisable, even as a mere verbal description. When we come to repeat his investigations, even more do the similarities leap out at us, even more are they utterly unmissable. It is hard to imagine someone else reading his paper and not seeing them, too.

But, if you have never seen (never noticed?) these forms and types of motion before, and have heard a relentless voice in the background telling you every day that they do not, can not exist, maybe it is explicable that you do not even notice them, cannot let yourself notice them, and therefore do not even notice a verbal description. The words are invisible to you. Years of training stop one seeing certain things. Training in mechanistic science demands such a crushing narrowing of one's inner vision, if it has not already been narrowed,[28] that it is no surprise that few readers or observers remain able to see Brown's words on pulsatory motion and the birth of living matter from dead.

An experienced paediatrician can examine a baby whose physiological and emotional integrity has been shattered by armouring and tell his parents that he is a normal healthy infant. Midwives and obstetricians witness the effects of muscular armouring every day in childbirth and do not see it. Psychiatrists treat depressed people every day and do not notice their inhibited breathing. So why should biologists be any more observant than these and notice the bions under their microscopes?

Addendum, late 2010.

Yet again, by coincidence, I have come across another example of this relentless drip, drip, drip of deception, a video on YouTube from a BBC documentary on the history of nuclear physics, broadcast in 2007, *The*

Atom. I came across this film after reading firstly Isaacson's biography of Einstein,[29] which contained much material about the German nuclear physicists during World War II, and secondly *Hitler's Scientists*,[30] which filled in the story for me to some extent. After reading this book I wanted to know more about Werner Heisenberg, the leading German nuclear scientist, as a person, not only about his scientific achievements. I put his name into a YouTube search and this documentary appeared. I was not looking for further material and proof of scientific deception. It came and got me.

In the first episode[31] the presenter, Professor Jim Al-Khalili, professor of nuclear physics at Surrey University, gives viewers a potted history of Brown's discovery. He, too, claims that Brown sprinkled pollen grains on water, (a most unlikely way of carrying out his experiment, given the size of his microscope stage), and implies that he immediately noticed what we now call Brownian motion. We see Professor Al-Khalili sprinkling something on some water, he looks down a modern binocular microscope, and then we are shown, apparently, what he saw. Now if he had really sprinkled pollen grains on the water, he would not have seen what the camera shows us. For a start there would have been some much larger items visible, the actual pollen sacs themselves, from which the pollen grains emerge as the sacs rupture in the water. As we view the Brownian motion of the 'pollen grains' in the documentary, there is an image in the background of what are undoubtedly pollen sacs, lumpy little forms with a grainy surface. But these pollen sacs do not show Brownian motion. They are far too large for that. The graphics here are little better than those of the *Usborne Dictionary of Science*. There is no actual mention of or quotations from the text of Brown's original paper.

Once more the innocent viewer goes away without ever suspecting that Brown had described different forms of motion from those described by the professor, pulsatory forms that led Brown to use the word *vermicular* about some of them, and that he originally thought he had discovered the primary elements of life, which he named *active molecules*.

---

1 Einstein A and Infeld L (no date); The Evolution of Physics, Scientific Book Club, London. This book was first published in 1938. It has been published in various editions since then, including one as recently as 2008.

2 De Buffon (1781); Natural History, Volume II, Chapter VIII, Reflections on the preceding Experiments, Strahan and Cadell, London.

[3] Needham T (1748); A Summary of Some late Observations…Phil. Trans. Roy. Soc., volume 45, pages 615-667, London.

[4] Einstein A and Infeld L; op cit, pages 63-67 in edition cited.

[5] See page 119.

[6] Isaacson W (2008); Einstein – His Life and Universe, pages 463-465, Pocket Books, London.

[7] Brown R (1827); unpublished); manuscript note in slip no 24/239, Botanical Library, Natural History Museum, London. Thanks to library staff Natalie Pope and Armando Mendez for assistance in locating this item.

[8] Van der Pas P W (1971); The Discovery of the Brownian Motion, *Scientarum Historia* 13: pages 27-35.

[9] ibid, page 27.

[10] ibid, page 28.

[11] CB; chapter II, 2, The Question of "Brownian Movement".

[12] Strick J (2000); Sparks of Life, page 89, Harvard University Press, Cambridge (US) and London.

[13] Adams G (1787); Essays on the Microscope, cited in Gerlach D (2009); Geschichte der Mikroskopie, page 151, Verlag Harri Deutsch, Frankfurt am Main.

[14] Sloan P (1992); Organic Molecules Revisited, in Gayon J (ed); *Buffon 88, Proceedings of the International Buffon Conference*, pages 415-439, Vrin, Paris.

[15] Ford B J (2009); Charles Darwin and Robert Brown – their microscopes and the microscopic image, *Infocus*, Issue 15, September, 2009, Oxford. There is also a film of Ford's project available via Ford's website – www.brianjford.com and on YouTube. Type 'Brian J Ford' into the search window.

[16] Ingenhousz J (1789); Remarks on the Use of the Microscope, attached to van der Pas's paper, reference 8, above.

[17] Mabberley D J (1985); Jupiter Botanicus – Robert Brown of the British Museum, J Cramer, Braunschweig and British Museum (Natural History), London.

[18] Ford B J (1985); Single Lens – The Story of the Single Microscope, Harper and Rowe, New York.

[19] Ford B J (2009); op cit.

[20] Ford B J (1992); Brownian Movement in Clarkia Pollen: a Reprise of the First Observations, *The Microscope,* **40** (4): pages 235-241.

[21] 'Big Al' (2008); BBC – h2g2, Brownian motion, BBC, on line.

[22] Hartley W G (1993); The Light Microscope Its Use and Devlopment, page 38, Senecio Publishing Company, Oxford.

[23] Bowley R (2010); http://www.youtube.com/watch?v=FAdxd2Iv-UA

[24] Hart-Davies R (editor) (2009); Science – the Definitive Visual Guide, page 128, Dorling Kindersley, London.

[25] Stockley C et al (editors) (2007); Usborne Illustrated Dictionary of Science, page 123, Usborne Publishing, London.

[26] Kremer B P (2010); Das grosse Kosmos-Buch der Mikroskopie, page 27, Kosmos Verlag, Stuttgart.

[27] ibid.

[28] Hudson L (1967); Contrary Imaginations, Penguin Books, Harmondsworth.

and

(1968); Frames of Mind, Methuen, London.

[29] Isaacson W (2008); op cit.

[30] Cornwell J (2004); Hitler's Scientists, Penguin Books, London.

[31] http://www.youtube.com/watch?v=EQy_-wDRlk0 BBC 4 (2007); The Atom Part 1, Clash of the Titans.

Chapter 7 Repeating Reich's and Brown's Experiments

Few readers will think they have the facilities or the wish to actually carry out these experiments themselves. Please think again. The world of scientific orgonomy and microscopy is enthralling and raises one's study of orgonomy to a completely different plane. It is one thing to have read Reich's books, quite another to have repeated some of his experiments or to have built an orgone accumulator and to know from experience that his findings are accurate and repeatable. Another important drawback to the armchair study of Reich's books is that the photographs in *The Bion Experiments*[1] and *The Cancer Biopathy*[2] are almost meaningless to someone with no experience of viewing things down a microscope. I certainly found them so, until I had done the bion experiments myself.

The argument of this book is that Brown was, unknowingly, describing bions, when he reported on his investigations of his active molecules. To demonstrate this point we need, therefore, to carry out some basic bion experiments so that we become quite familiar with the bions and how they behave and can recognise them easily. We also need to be able to see if a preparation contains no bions at all.

If you are a complete beginner and have no-one to help you out with your first steps, a very enjoyable way of learning the ropes with your microscope is to collect a few samples of standing water from different sites and to examine them under low and intermediate magnification. In most samples you will find an interesting and beautiful sea of moving life before your eyes. This is far more interesting than looking at immobile items such as a human hair or the wing of a dead fly. There is, however, a good use for a slide of the wing of a fly, if you have one, or a similar item. If you wish to check the optics of a borrowed instrument or simply that your microscope is set up properly and producing a good image, it is easy to do this by looking at some item with clear, thin lines, so that you can see if you are obtaining good definition. Such ready-made slides are available from Brunel or Philip Harris and on ebay. (See suppliers' addresses, page 140-143.)

Another item which helps you to do this, and which is useful anyway, is a stage micrometer. This is a microscope slide with a diagram on it like the divisions on an old-fashioned ruler, except that the divisions are so small that you can barely see the diagram with the naked eye. You place this item on your microscope stage and see the 'ruler' magnified so that you can easily see the divisions. You should be able to get these into clear, sharp focus without difficulty, if your microscope is working well.

What do we need to repeat a basic bion experiment? The ideal tool is a research-grade biological microscope, that is, the sort of instrument that you will find in a university or hospital laboratory, probably with dark-field and phase-contrast facility. Such microscopes usually cost several thousand pounds and I do not suppose for a minute that any readers will at this stage even think of purchasing such an instrument. (You can buy an obsolete second-hand research-grade microscope for a few hundred pounds on *ebay*.) A basic student's model will be adequate for your purposes at present. This should be a binocular microscope with magnification up to 1000x and if possible should have phase-contrast facility and dark-field lumination, too. These types of lighting allow us to see things we would not normally see with simple lighting, that is, so-called bright-field illumination. Many student models have only bright-field lighting. We can see bions and bionous movement under such lighting, so such an instrument will be adequate, though it may make a beginner's work a little more difficult.

Where will you find such an instrument, if you do not wish to buy one? If you are lucky, you may know a professional microscopist or serious amateur who has their own instrument at home, and he or she may be willing to let you work on it, or better still, to let you borrow it for a while. Unfortunately there are not many private microscopes out in the world. Apart from those in the home labs of orgonomists in America and Greece, I have not come across a microscope in anyone's home. It would have been a common item in many middle-class drawing rooms in Victorian England.

However, there are many unused microscopes gathering dust in laboratories and colleges. If you know someone who works in such a place, you may be able to wangle a loan, better still, a bargain purchase. Help from an experienced user who can show you how to use it would be useful at this point. However, an enterprising learner can work out how to use a microscope just by looking at it and slowly and carefully trying it out step by step. If you are offered one of these allegedly obsolete instruments from a lab where they have recently bought new models, take it with both hands. You can always work out how to use it, given time. If you look at a microscope, it is fairly obvious where to put your hands and which bit does what. The microscope is a much kinder piece of technology than a computer or motor-car and if you move carefully you cannot do too much damage. There is no such thing as a 'crash' on a microscope. The only major harm that you could cause is, as far as I can see, damage to an objective by absent-mindedly winding up the stage with the coarse-focus wheel without the stop-lever set correctly, grinding the objective against the slide. If you do this on a high-quality microscope the objective could cost several hundred pounds to replace, possibly even more. Even on a cheaper model it could cost £200 to replace a 100x objective. If you crunch a slide to bits, the glass fragments may get into the condenser and do harm there, too. Above all, particularly while you are learning to handle your instrument, move slowly and gently. At all times move slowly and gently while using a microscope.

I shall suppose that you have somehow got your hands on a usable instrument. What else do we need? Next to nothing! Only a few simple extras, most of which will also be available from the same place that you have borrowed your microscope from. These sundries will almost certainly be available anywhere where there is a microscope, as a biological microscope is useless without them. If you happen to have found an unused, long-lost instrument gathering dust in an attic or basement, or have bought yourself a cheaper student's model, than you will need to buy these extras. A rough calculation from Brunel's on-line catalogue puts their price at about £50. Items marked* are absolute musts. The others can be improvised.

These are:

- Test-tubes*
- Stoppers for the tubes
- Test-tube holders
- Test-tube rack
- Microscope slides*
- Coverslips*
- Coverslip forceps*
- Sterile swabs (*Sterets* or *Mediswabs* available from chemists)
- Pestle and mortar
- Spirit stove, bunsen burner, or some form of open flame
- Syringes and hypodermic needles*
- Cotton-wool balls
- Oil for oil-immersion observation*

Important: Safe Disposal of Sharps

Before you start any of this work you must devise a way of collecting all your 'sharps' and eventually of disposing of them safely. This word may be new to you; the word 'sharps' is used in hospitals and laboratories to refer to any item that may potentially cut someone and which may be infected. All your coverslips, any broken microscope slides, and especially any needles that you use to draw up samples from your bion preparations are sharps in this sense, even though they are almost certainly not contaminated with anything that could cause a serious infection. A council refuse disposal worker will not know this, if they find a needle in your domestic rubbish. If they are stabbed by it they will be panic-stricken, thinking, quite reasonably, that the needle has been used by a drug-user and that it may well be contaminated with hepatitis germs or HIV. Hepatitis is the greater danger, though HIV frightens people more than hepatitis. Most needles found lying about in refuse have been dumped by drug-users, so the worker concerned has reason to think that your lab needles come from the same stable. Sharps are disposed of in hospitals and labs by incineration. They are collected in tamper-proof plastic containers and the containers are thrown into incinerators without anyone having to risk injury. You can buy small sharps boxes from science suppliers, quite adequate for this work, as you will not be discarding any large items. You can also improvise a sharps box by labelling a plastic milk-bottle and keeping the lid to hand, securing it when it is not in use. You still need to find a safe disposal point. If you know someone who works in a GP's surgery or hospital or lab they may be willing to dispose of it safely for you. Otherwise contact your local council for advice. Most councils have needle-collection points for drug-users.

It is a positive exercise to draw what you observe under your microscope. An easy short-cut is, of course, to use the ubiquitous digital technology and just photograph or film your slides, though this will do nothing for your powers of observation. Drawing makes you really look at what you are observing. To do this you need some artist's drawing paper and a few good pencils of varying hardness and softness, say, B, 2B, 3B.

Many of these items are found in homes and some can be improvised. The items marked with an asterisk, as far as I can see, cannot be improvised and will have to be bought. Bear in mind, though, that the first microscopists must have managed without these items and so it must be possible to make some of these observations with very little equipment indeed. All these items are fairly cheap. For instance, a box of a hundred test-tubes will last you for years and cost only a few pounds, as will a box of a hundred slides. Plastic syringes and sterile needles also cost only a pound or two for ten and are available from educational suppliers. £50 should keep you going for a long time, once you have got a microscope.

Just in case you happen to have a microscope with no-one to show you how to use it, I append a picture with the names of the basic parts and some information on what they do. The parts marked in the picture are printed in **bold type** in the explanations below. This picture is modelled on a Brunel SP100 model. You should be able to adapt it to any other microscope, as biological microscopes are all more or less similar.

As readers can see, none of the names of the parts are surprising or obscure. Even if you do not know these names, you can still work out what the parts do and how to handle them. A basic microscope is scarcely more complex than a camera and no more difficult to operate. (A microscope has in fact a lot in common with a camera.) For further information on how to set up your microscope, please refer to one of the textbooks cited at the end of the chapter or C O R E's booklets on the microscope.

The **eye-pieces** are the parts that we look down. Most eye-pieces magnify by 10, producing a maximum magnification of 1000x, (the objective's 100x multiplied by the eye-piece's 10x = 1000x). An easy way of getting much higher magnification is by using eyepieces with higher magnification. 20x and even 25x eyepieces are available. If you are thinking of buying a pair of these eyepieces check the diameter of the tubes that they must fit into. There are two standard tube-diameters for eyepieces, 23 mm and 30mm. Most microscopes use the 23mm-diameter eyepiece. A high-magnification objective, above 100x, gives a better image than endless magnification upwards by the eye-pieces, but these are now very rare and, as far as I know, are not manufactured any more. Therefore 20x and 25x eye-pieces, are the next-best option. Some older microscopes have built in magnification changers, (An East-German Zeiss we bought on ebay includes one with 1.6x magnification and C O R E's Olympus has a 2x changer which has been added recently.) These are a great convenience. They mean you have at your disposal high magnification without using oil-immersion. (You cannot go back to low magnification, which you may need to do now and again to find something on a slide, once you have got oil on the coverslip.)

The **stage** is the movable platform on which the object to be viewed rests. A stage that can be moved from side to side and forwards and backwards by the operation of wheels turned by the hand of the viewer is known as a **mechanical stage**. Such a stage is a necessity for these experiments. Some cheap beginners' microscopes have a stationary stage with a pair of spring-metal prongs that grip the slide and hold it in place. You have to move the slide with your fingers as you observe. This will make life very difficult for a beginner and it will be even harder to relocate something that you have found and wish to come back to after looking at something else.

The main optical component that gives most of our magnification is known as an **objective** and is the tubular metal item just above the stage. There are usually at least three of these on most biological microscopes, screwed into the nosepiece, which rotates manually into position above the stage.

The **condenser** is the item below the stage and contains the optical components that bring the light to a coherent beam for the purposes of the viewer and the lumination required. Some condensers are quite simple, if there is only one type of lighting available on the microscope, and may consist of a simple lens and an iris diaphragm that increases or decreases the aperture and so the diameter of the light-beam shining upwards through the condenser and into the objective. This diaphragm is operated by a simple lever. More complex condensers may possess facilities for phase contrast and darkfield illumination.

The **illuminator** or **light source** controls the intensity and diameter of the light beam. It may have no controls at all on a very cheap microscope. More expensive instruments will include a rheostat adjuster (a slider or rotating knob) for the light intensity and better models also have a field diaphragm which adjusts the diameter of the visible light beam. This field iris is an essential item to obtain full Köhler illumination. (See glossary.)

The **focus wheels**, **coarse** and **fine** are positioned together concentrically, as in the picture. The coarse-focus wheel winds the stage up or down much faster than the fine-focus wheel and should be used carefully and sparingly. All better instruments and many quite basic ones have a **stop lever** that prevents you winding up the stage beyond the point at which the lever is set. To wind up the stage beyond this point you use the fine-focus wheel, which is much slower and avoids the risk of a 'collision' between the slide and objective. The instrument illustrated has a stop lever.

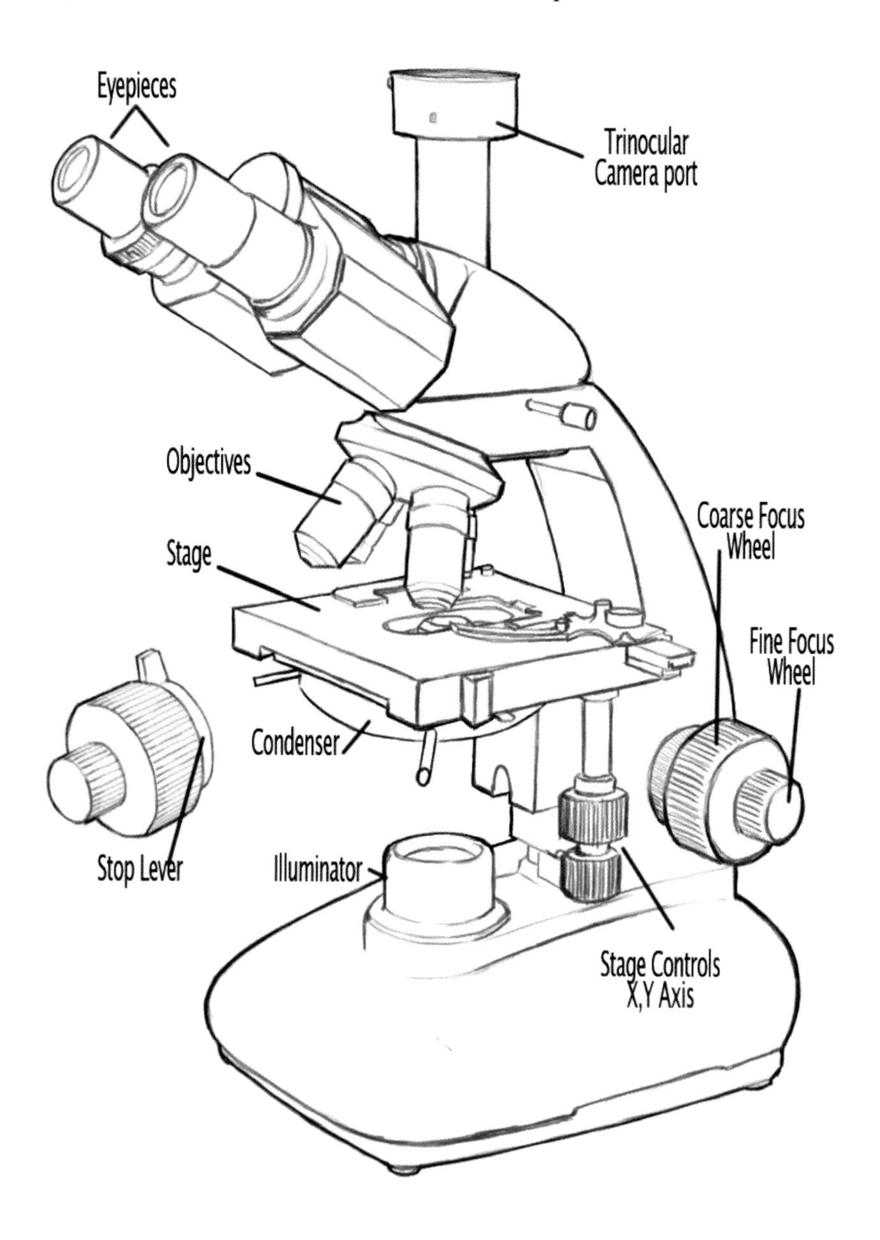

A **slide** is a small sheet of glass, typically, 25mm x 75mm, (1” x 3”). The specimen or the object that you wish to examine is placed in the centre of this glass and covered with a **coverslip**. This is another small piece of glass that is very thin indeed, typically 22mm x 22mm. We pick this up with a pair of **coverslip forceps**. These allow us to handle the coverslip safely and to place it exactly where we want it on the slide. These slips can be bought in various thicknesses. In practice there is only one commonly used thickness, known as 1½.

A **well-slide,** sometimes called a cavity-slide**,** is a similar item to the above, but it has a small depression in the centre, which allows us to view a hanging-drop slide. With one of these, the item to be viewed, a drop of fluid, is suspended from the coverslip, hanging down into the well.

A Basic Bion Experiment

I call this a basic experiment, because the techniques and practical considerations and problems are all the same whatever you decide to grind up and use in your experiment. I choose ground red clay as this is cheap, commonly available and completely safe to handle. To obtain my clay I located a craft potter in the local Yellow Pages, rang them up, and asked if I could buy a pound of clay. (Asking a potter to sell you a pound of clay is a bit like going to a grocer’s and asking for a teaspoonful of milk. Potters use clay by the hundredweight, but this one was quite happy to let me have a pound of clay for a few pence. I explained that it was for educational purposes.) Another very amenable material is the rock-dust sold to revive exhausted soils. This is sold by horticultural suppliers, mainly in large quantities, though you can purchase 340g bags from some suppliers.• This is mainly basalt in origin and is already coarse dust when you buy it. It can easily be further ground to produce a nice, smooth fine-grained sand that is very easy to handle and which produces generous bion growth very reliably. You could also use coal-dust, sand, soil, ash, slate, brick, any stone available locally, iron-filings, other sorts of metal filings available locally, bone-meal, ground rice, Fuller’s earth, fossils, vermiculite granules,

• See the website of the pioneers of this medium, www.seercentre.org.uk for a list of suppliers and information, SEER (Sustainable Ecological Earth Regeneration) Centre Trust, Ceanghline, Straloch Farm, Enochdhu, Blairgowrie, PH10 7PJ, Scotland.

chalk, or drawing pastels, and anything else that you can find. It is an easy matter to make iron-filings or other metal filings by simply rubbing a coarse file against a bit of scrap-iron or a copper washer and catching the dust as it falls. You can buy small samples of almost any conceivable rock or mineral from geological suppliers. As you will see later, we need some rather out-of-the way items to check Brown's findings, including a sample of fossilised wood. This, too, is available from geological suppliers.

Remember that 'sand' is only rock ground up by natural processes and varies in content according to the local rock. If you get seriously involved in bion research it is interesting to test sand from different locations and even to make your own 'sand' by grinding finely some material such as sea-shells, egg-shells, crab-shells, local rock, and so on.

So… if you are using some other common material, just replace clay with your item as you read.

Preparing materials

It is safest to use inorganic material with no combustible contents. When you come to sterilise your materials in a flame, anything containing organic matter will spit and pop and possibly catapult small blobs of very hot or burning material out onto you or the nearest surface or the floor. This can be dangerous and painful, needless to say. I suggest you leave soil until you are confident and experienced and know how to handle hot materials. If you are working in an improvised laboratory with no proper bench and fire-proof measures, heat your test tubes and materials on spatulas over the metal surface of a cooker. That at least cannot catch fire, if you drop some red-hot material on it. It is best and safest to buy a mortar and pestle and use them only for your bion experiments. One used for cooking purposes will be contaminated with food particles, which will burn when heated.

Let your clay or chosen material dry out, pass it through a strainer, so that there are no lumps in it, and grind it up as finely as you can. If you are using a spirit burner to boil your tubes, have it ready to light and in a safe position clear of items that might get in the way or cause fire risks. Have a test tube ready in a rack or improvised substitute with a suitable stopper to hand. Simmer about 10 mls of water in the test tube until it is boiling gently and evenly and let the boiling fluid rise up the inside of the tube, so that it has covered the internal surfaces and sterilised the inside fully. Discard this water. Add a similar amount to your tube and bring it carefully to the boil again. Seal your tube temporarily with a cotton-wool

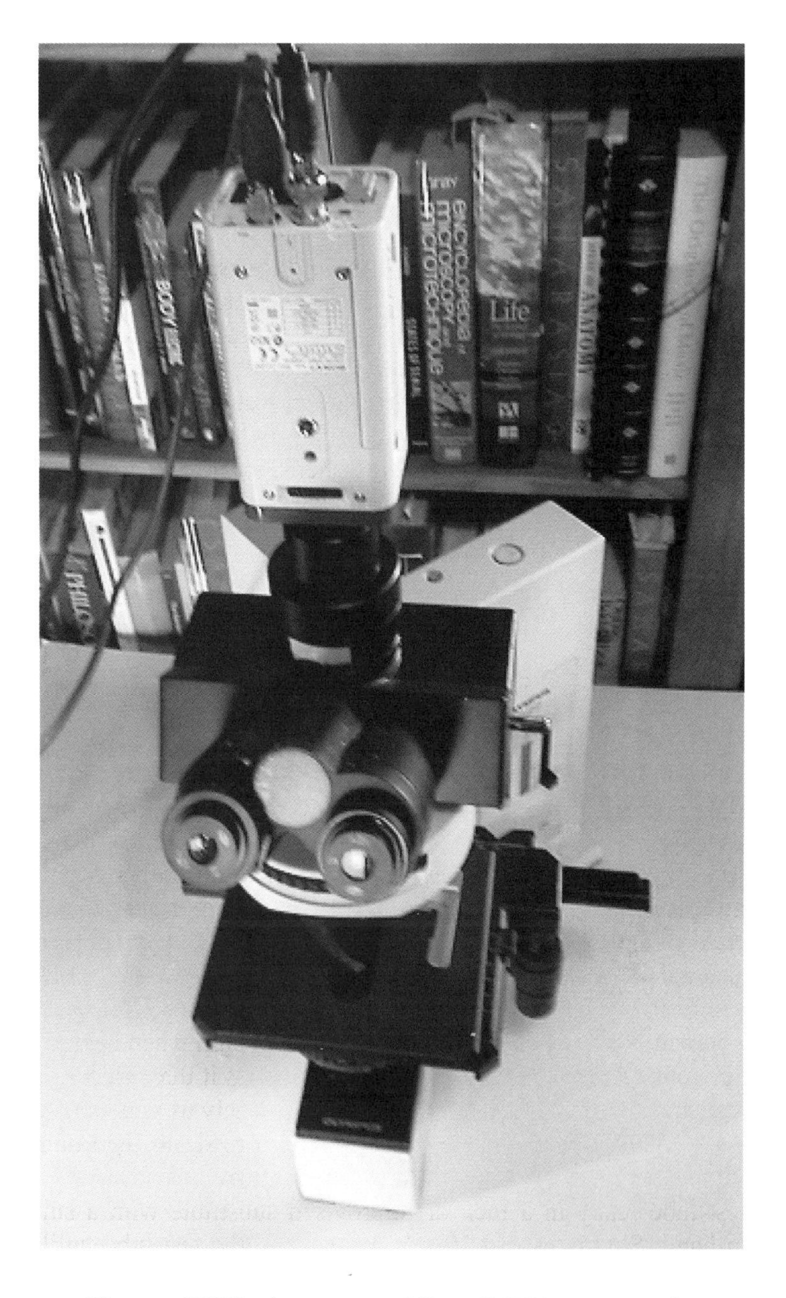

Olympus BX50 microscope and Sony CCTV camera used to film the experiments described in this book.

Robert Brown's microscope. (By permission of the Linnean Society of London.)

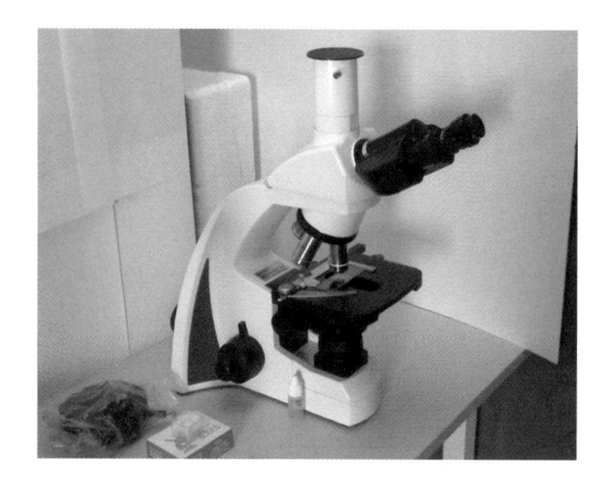

Brunel SP150 microscope in C O R E's improvised lab.

It can be done! A Brunel SP100 Microscope on the corner of a table with instruments and sundries. The 'bottle of milk' is an improvised sharps container. All the experiments described in chapter 7 can be done with this microscope.

Wilhelm Reich
(1897-1957).

Robert Brown
(1773-1858).
By permission of the
Linnean Society of
London.

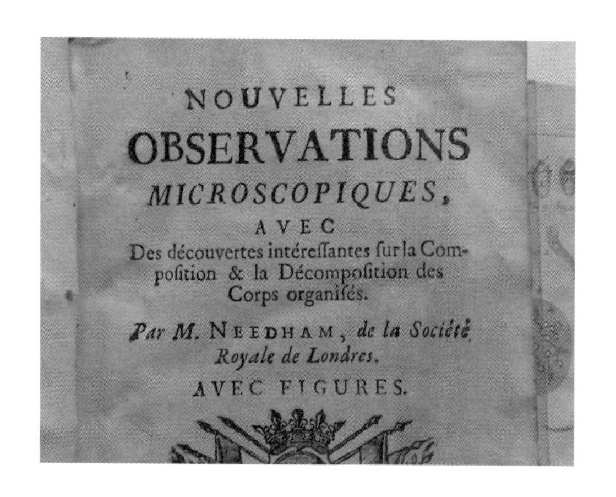

NOUVELLES
OBSERVATIONS
MICROSCOPIQUES,
AVEC
Des découvertes intéressantes sur la Composition & la Décomposition des Corps organisés.

Par M. NEEDHAM, de la Société Royale de Londres.

AVEC FIGURES.

Frontispiece, Needham's *Nouvelles Obsercations Microscopiques* (1750)

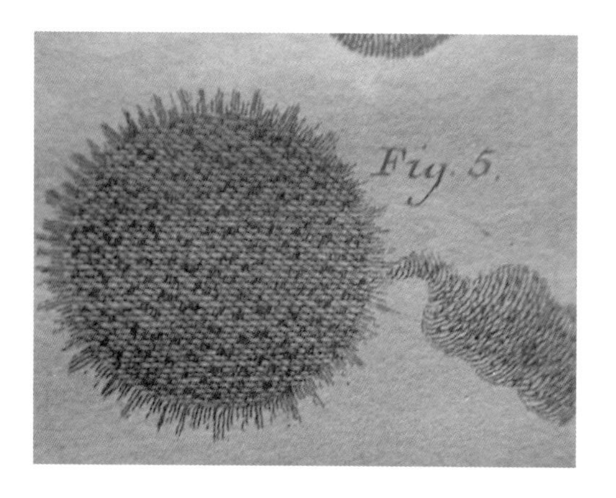

Needham's illustration of pollen grains emerging from a pollen sac in water (1750). Note similarity to picture on next page.

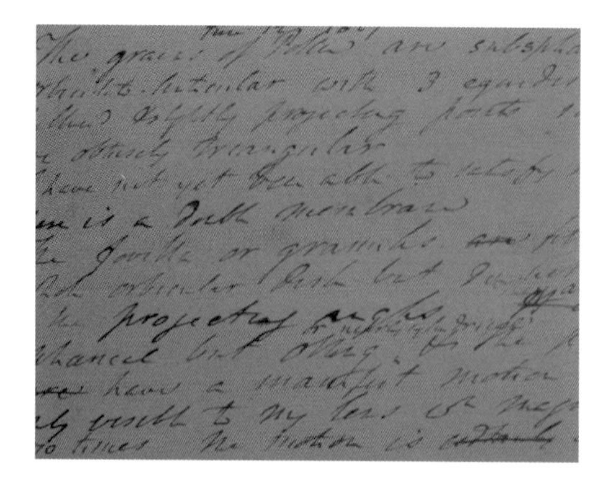

Sample sheet of Brown's slips, the original pages of his botanical notes. (Photo courtesy the Natural History Museum, London)

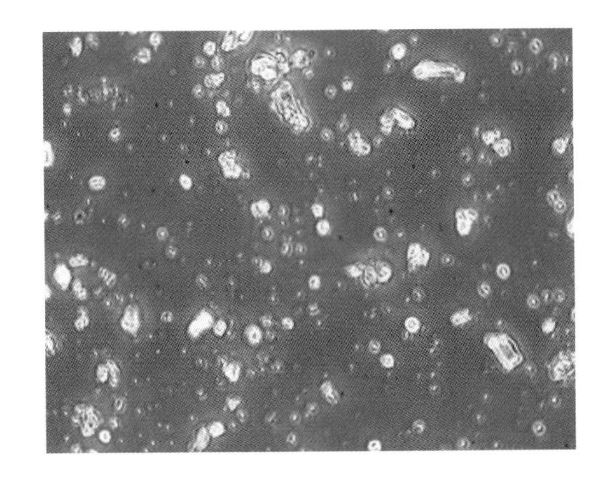

Bion Culture made from rock-dust heated to red-heat prior to addition of fluid. The bions are the smaller black dots visible here and there. See C O R E's video on YouTube to observe their bionous motility.

A hay-infusion, boiled 10 times and autoclaved. The formation just off centre is contracting rhythmically and rotating on its own axis. See YouTube video to observe this motion This form is described by Brown in his 1828 paper.

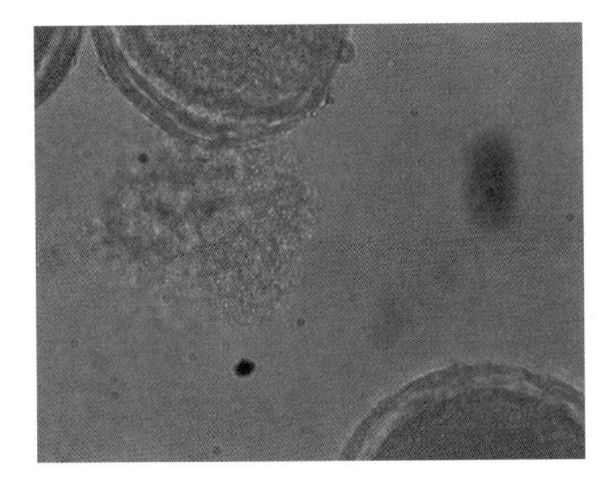

Bionous granular matter bursting out from a freesia pollen sac under water. Already a few bions are dancing about here and there.

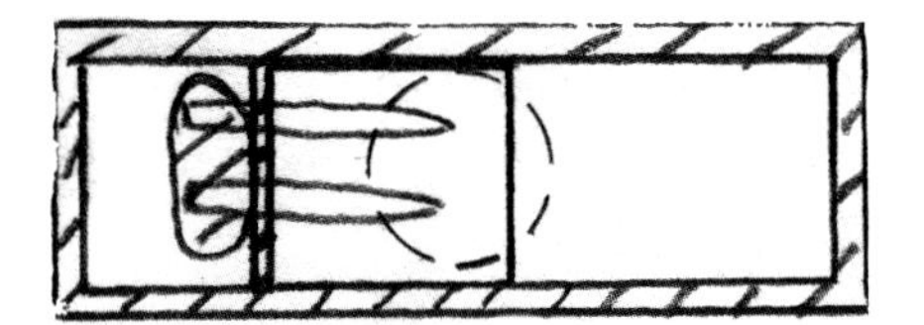

Grass-infusion slide. Hatched areas are wax mixture.

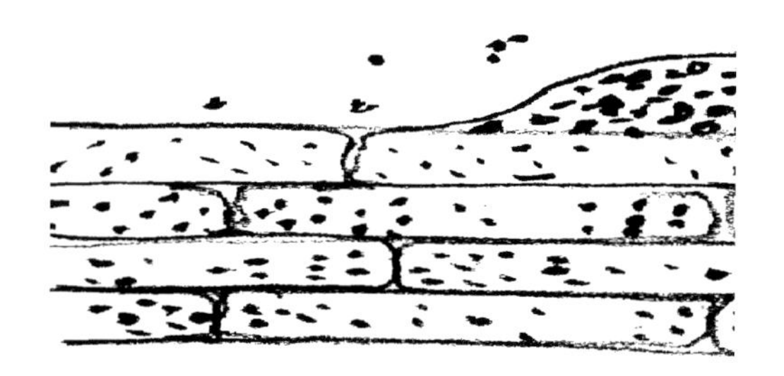

A typical swelling on the edge of a grass-blade with motile bions inside itand free-floating bions above the surface of the grass.

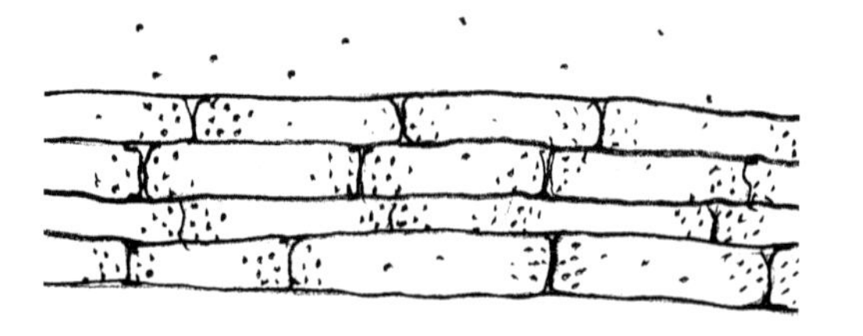

Free-floating bions in water and bionous disintegration starting on grass surface around cell borders. Drawn from life, simplified.

ball and place it in your rack or improvised substitute. Scoop up a spatulaful of your ground material, (if you have no laboratory spatula, use an old teaspoon), and place it in the flame of your gas cooker or spirit burner. Move the spatula or spoon to and fro in the flame, making sure that the little pile of material gets evenly heated. At least the edges should glow red and ideally the whole amount should. A spoon is thicker than a laboratory spatula and will absorb quite a lot of the heat of the flame and you will probably be able to get only the edges of the material to glow red. The rest of the pile will certainly be hot enough to exclude the survival of any possible germs, even if it does not glow. (If your spatula or spoon-handle gets too hot for comfort, hold it with an insulating cloth.) The temperature of material heated to red heat is, according to a technical dictionary in C O R E's library, between 500°C and 1000°C,[3] so presumably material close to other materials that are glowing red is at least at 200-300°C, easily hot enough to destroy any living organisms within your ground up solid. When your clay has glowed for a few minutes, tip it off the spatula into the test tube. You will need to add about three spatulafuls of clay or dust. (Here a spatula means one of the narrower, lighter sort, about 2-3 millimetres wide, not the wider, heavier type.) Place a cotton-wool bung in the neck of the tube or close it with a cork which has been cleaned with an alcohol swab. Your bion preparation is now brewing, though one made with ground clay or rock-dust needs very little time to produce generous numbers of bions. You can draw up a sample with a syringe and examine it straight away.

Note on Brownian Motion

While re-reading and revising this chapter it dawned on me that I was continually referring to Brownian motion and bionous motility and pointing out that the motion seen in the bions is not Brownian motion. Beginners may be asking; what then *is* Brownian motion? (See this reference[4] or any physics text-book for a conventional definition.) If you have not yet found something that demonstrates it clearly to you, imagine a tiny particle being buffeted from different directions. It will jerk first on one direction and then in another, including 'up' and 'down', that is, towards you as you look down your microscope and away from you as well as to left and right and up and down on the slide. While you are getting on with your bion experiments, you should be investigating a good few possible fluids that may show Brownian motion clearly. Once you have seen it clearly, you will be in no doubt as to what is and isn't Brownian motion. Learning to recognise pure Brownian motion is a basic part of your education in this field. Please don't neglect this. You cannot really make any reasonable claims in favour of bionous motility until you are sure of the difference between these two forms of motion. There are films of 'Brownian motion' on YouTube, but, significantly, most of them are either animated models or smoke particles in air. The usual, neutered version of Brown's paper accompanies some of the films. (Notice: I deliberately have not named any fluids that simply show Brownian motion. The investigative process involved in the search for Brownian motion will teach you a great deal. You will know it when you find it. You will almost certainly be able to observe Brownian motion in some item within your own home.)

The Move to the Microscope

If you are a beginner, at this stage you need to be sure that your microscope is ready for use, connected to the power, adjusted as needed, and so on. Once you have set up a few slides, you will do all these jobs without thinking and, anyway, your microscope will be more or less ready for use all the time, if you own one and have a fixed home for it. If you have borrowed one and have negotiated a temporary home for it in your household, such as corner of a dining table or work-bench, life will be more difficult. You will just have to spend more time getting things ready and finding all your items. This will be frustrating at times, but not impossible. Even at C O R E, after twelve years' microscope work, we still have not got running water in our 'laboratory'.

If you are using sterile syringes and needles directly from their packaging this stage is simple. Have a slide, coverslips, and coverslip forceps to hand. Remove your syringe from the packing, place one of your paediatric size needles on the syringe, give your tube a gentle shake to stir up the particles in the fluid, remove the stopper, insert the syringe into the fluid carefully and draw up a small amount of the preparation. 0.2-3ml is plenty, although you will soon find yourself judging this by eye.

At this point we need a coverslip to cover our sample on the slide. Coverslips are awkward to handle until you get used to them and need care. If you drop one, it is often impossible to find it. It then gets trodden on and can injure anyone who treads on it with bare feet or who is cleaning or picking something up with their hands. So, be careful here. I have marked coverslip forceps as absolutely essential and do not recommend trying to set up a slide without them. To place a coverslip over a fluid sample on a slide, I extract with the thumb and fingers of my left hand a single slip from the top of the pile in their plastic container. After a little practice you immediately realise if you have picked up more than a single slip with your finger and thumb. Another way of checking this is to stroke the edge of the slip with your finger and thumb, once you have picked it up with the forceps. A single slip has a characteristic bendy feel and also makes a particular sound which I could not possibly describe, but which you will soon recognise. Two coverslips used by accident will make problems for you if you are using an oil immersion objective with a very small working distance. If you want to be extra careful with your sterile precautions you can 'flame' your slide and slip by passing them through the flame of your burner with the forceps before you use them.

Drop three to four drops of your preparation onto the centre of your slide and carefully place a coverslip over the fluid. It should spread out tidily under the coverslip, so that you now have a thin film of your preparation all ready to be examined under your microscope. If you have put too much fluid on the slide, the coverslip will float on it and cavort about the slide. You can reduce the amount of fluid by soaking up some of it with a little strip of tissue or start all over again. In very hot weather the surplus will evaporate before long anyway.

How do we get the slide into focus so that we can examine our culture easily without strain or difficulty? If you have already familiarised yourself with your microscope by examining a few samples of pond-water, this should now be straightforward. You should get to this stage before trying to examine bions. This is fairly easy to do and demands only a modicum of care and concentration. Place the slide on the stage in between the spring-loaded clips that hold it in position. I assume that you already know how to find without looking the focus wheel and the controls of the mechanical stage. (You can't look at them because you are looking down the eye-pieces.) Start with the low-magnification objective in position.

If you are starting with a 10x objective you will need to close down the aperture in your condenser and the field diaphragm, if there is one, in your illuminator. The rheostat setting will also need to be fairly low. (You should already be familiar with the appropriate settings from your observation of water life.) Wind up the stage with the coarse-focus wheel until you can see blurred shapes coming into view down the eye-pieces. At this point set the stop-lever and wind up using the fine-focus wheel. The coarse-focus wheel won't go any further once you have set the lever and this is a safety feature, even on fairly cheap binocular microscopes, to stop you winding up without looking and grinding your objective against the slide. With the fine-focus wheel you will feel or hear a gentle click if you wind up too high and bump into the objective. As you wind up, watch carefully and notice when the items on the stage come into focus.

It is a good habit to get into, and anyway an effective way of learning to use your instrument, to start with the lowest magnification and move upwards in steps. So let us begin with the 10x objective and bright-field lighting. The 10x objective will probably be the smallest of the three on your microscope. If your instrument has an even smaller one, say, 4x, you will have an even lower magnification at your disposal. At 40x (10x x 4x) the smaller particles of a clay or rock-dust preparation will be barely visible. At 100x the large particles will be recognisable as such. Look carefully between these larger particles. If you are observant, you will actually be able to see tiny dots, so small as to be barely visible, dancing before your eyes. These are probably bions.

When you feel you have seen everything you can see at this low magnification, move up to intermediate magnification with your 40x objective, which will probably be the next larger one on your microscope. On C O R E's Olympus the 40x objective is a phase contrast one and we also need to move in the corresponding phase contrast component in the condenser. (A phase-contrast objective can be used without its phase-ring in the condenser and will then just give you a normal brightfield image.) This allows us to see some forms that we would not otherwise see, though with a simple clay culture it is not too helpful. At 400x we can see much more and what we may eventually recognise as bions are quite easily seen at this level. The large crystalline clay particles are immediately recognisable. So, where are the bions? Look carefully at the space between the particles. Now look at the highly motile 'dots'. Observe these small particles very carefully indeed.

Many of them will certainly be highly motile. If their motion is Brownian, all the particles of a certain size should be in motion. (The size of particles in motion is crucial to Brownian motion. Once particles become larger than a certain size, they do not show Brownian motion.) You will almost certainly find some tiny dots in motion and some of a similar size that are not moving. You will almost certainly see pairs of these dots cavorting around each other, as if there were some invisible attraction between them. (If we repeat this experiment with different solid materials, we may even see bions 'dying'.) You may also see an agglomeration of bions that move about of their own accord and even bend and contract like a living being. This characteristic movement was described by Brown in his *Active Molecules* paper. If you observe for long periods, you may be lucky enough to see one of these agglomerations forming. You may also see a pair of bions linked by what Reich called a *radiating bridge*.[5]

To see more detail and be sure of what we are observing we need to move up to the maximum magnification on our instrument, 1000x (10x x 100x). We need to use oil immersion for this. Don't be put off by apparently fancy words. This technique is quite simple. Swing out the 40x objective so there is space above the coverslip. Wind down the stage of the microscope until there is enough room to insert the nozzle of the oil-bottle above the coverslip. Squeeze the oil-bottle gently until a drop of oil appears at the nozzle and drops off onto the centre of the coverslip. Swing the 100x objective into position and wind the stage up carefully until the lens of the 100x objective is almost touching the oil. Next wind it up very carefully using the fine-focus wheel until you see a flash of light as the oil and lens make contact. Then look through the eye-pieces and wind up slowly until items start to come into focus. You should then see your clay particles and, we hope, in between the larger particles, tiny spinning, dancing dots, the bions.

If you are looking at your slide using a cheaper microscope, you will now be at the maximum magnification possible on it. A possible option on some instruments is to use 15x or 20x eye-pieces, which will give you a magnification of 1500x or 2000x. (The Brunel SP40, SP100, and SP150 models have 20x eyepieces available.) These are definitely worth having. Just to show you how much easier the use of a high-quality instrument is, I must point out here that C O R E's Olympus accommodates a magnification-changer that doubles whatever magnification we get via the objectives and eyepieces. This means that with our 20x eyepieces and the changer we can get a dry magnification of 1600x using the 40x objective. This is very convenient and saves the interruption needed to set up oil immersion. It also means that you can go back to lower magnification to find something particular. Once you have got oil on your coverslip you cannot go back to dry magnification and so oil immersion is an all-or-nothing option.

If this first slide has not worked and you have not been able to recognise any of the forms that I have described here, I suggest you go back to some convenient point, say, drawing up a sample with your syringe, and start all over again. The actual culture is so simple that it is very unlikely that you have got that wrong. Check your microscope. It is possible that you have not got it set up correctly and that no light is getting through your slide and into the eye-pieces. It is helpful to be able to check your optical set-up by having a single prepared slide that you know well and which you can use as a test piece. You can buy a handful of such prepared slides from Brunel microscopes. Pick on an obvious one, a fly's wing, say, have a good look at it under all magnifications and conditions until you are confident that you can set up your microscope in a workable condition every time. Take a look at this slide before you start work to make sure that you are obtaining a sharp focussed image with the fine lines of the insect anatomy showing up clearly.

If you are getting no image or only a very poor one, it is possible that you have inadvertently left the condenser wound down. It normally needs to be wound up as far as it will go, right beneath the stage. It is also just possible, though unlikely, that the objective has not clicked into place properly and is slightly out of line. Check this by gently pushing it to and fro until it clicks decisively into place. Remember also that you will need more light as you increase your magnification. Open the condenser iris lever a little or turn up the light switch, if it is adjustable, or both. You should have played about with these controls before you start these experiments so that you are fairly confident with them and know at what levels to set them to get a good image.

These precautionary remarks are almost irrelevant. You will find that you can obtain a clear image almost every time you set up a slide. The microscope is a simple, reliable tool that almost always works, producing a good, clear image.

So…we have got to high magnification – 1000x, possibly 1500x or 2000x, if you are using 15x or 20x eye-pieces. What do we see? The larger immobile clay/rock-dust particles are now easily recognisable to you, I hope. Look carefully in the clear space between them. You should be able to see small dancing dots here and there. If you have used clay for your first bion culture, there should be these highly motile forms all over the place on your slide. Take a second, a third and even a fourth look. Let your eyes wander over the visible area. You should be able to find without too much difficulty a pair of these motile forms dancing round each other, moving to and fro, towards each other and away from each other. They will also be spinning in different planes, tumbling over and over themselves. The single ones moving without any apparent attachment or attraction to another particle will also be moving in this characteristic way. You can recognise this form of motion by ascertaining the shape of the moving particles. If they are at all asymmetrical you can see an ever-changing outline as they tumble over themselves all the time. Some of the larger motile particles will be rolling quite slowly.

I suggest that you try a few other materials and become quite familiar with the characteristic motion of bions before you try any of Brown's experiments. If you make a preparation from any material that produces a high pH you may be able to see bions 'dying'. If the pH of your preparation in water is above about 12.0 it may produce no bions at all. (This is not very likely. Very few materials produce such a high pH value.) If your pH value is borderline it may produce bions with a very short life and you may see their movement stopping. (Don't forget that this stopping of motion may just be because a bionous particle has sunk to the bottom of the layer of fluid on the slide and grounded on the glass of the slide.) Some preparations also produce very weak bions that only start moving very slowly if at all and you may be able to see this happening, too. Clearly this cannot be Brownian motion.

You can obtain 20x eye-pieces for all three Brunel instruments the SP40, SP100, and SP150. These eye-pieces give you an intermediate magnification without oil-immersion of 800x and will in fact work with most microscopes. This is high enough for you to see bionous motility and to make out the shapes and the mode of movement of the smaller particles on your slide. This fairly high intermediate magnification means that you do not have to use oil-immersion for your studies to start with and this saves a beginner a great deal of bother. I use this magnification a lot when I am doing routine investigations, repeating a familiar experiment, or just checking a finding. Obviously you will want to use oil-immersion and your high-level magnification eventually to learn as much about the bions and their ways of moving as you can before setting up the grass-infusion experiment, but the 800x obtained with the 20x eyepieces and 40x objective is a very convenient working magnification and quite adequate for general bion work.

If you are lucky enough to have access to a microscope with even higher magnification, in excess of 3000x, you may just be able to see the internal blue orgone energy charge and the internal orgonotic pulsation in your bions. C O R E's Olympus is equipped with a Leitz 160x objective. With the magnification changer and 20x eye-pieces this gives a magnification of 6400x. As far as I know, these very high-magnification objectives are no longer made by anyone and you will have to search for a second-hand one, if you wish to advance to this high-level work with the bions. Thanks to an alert orgonomic colleague, we found this marvellous objective on ebay. Unfortunately there is no way we can obtain these very high levels of magnification on the new, cheaper Chinese models as they have no provision for a magnification changer. Occasionally second-hand models for sale on ebay have a magnification changer. However, a 160x Leitz objective with 20x eyepieces on the Brunel SP150 gives excellent images at 3200x!

The Grass-Infusion Experiment

This experiment, although it is the foundation of Reich's bion work and came first in his own research,[6] is more difficult to do and I have, therefore, placed it after the basic bion experiment, so that readers come to it with some experience and after getting to know what bionous motility looks like. The difficulties are mainly physical. It is rather a fussy business to set up this infusion, which can be maintained continuously for a long period, though it is by no means impossible, even for a beginner with the microscope. It demands a certain agility and finesse with the fingers, not a degree in microscopy. We need some extra items to set up this infusion.

Items needed in addition to the list on page 74

Vaseline jelly
Liquid paraffin
An artist's paint brush made of natural bristles, not synthetic material
A heavy pan to melt the waxes in
A few blades of grass
Sharp scissors and safety razor
Cavity slide
A chopping board (or scrap wood) to cut the grass on

If you are at all practical, a glance at the picture (centre pages) tells you exactly what you need to do with the various items. It is a good idea to have a dummy run with your blades of grass to make sure that you can split them safely and accurately and get them into position on a slide. When you do this for the first time, I recommend you to do it on a day when you have got plenty of free time, with nothing pressing to do afterwards, so that you can work really carefully and slowly, giving the job all the care and time it needs. The infusion takes about 24-48 hours to produce bions and motile forms, so you also need to think ahead and plan a schedule that will allow you plenty of observing time at that point. You should also observe the infusion closely while it is brewing. The changing states of the grass and the appearance of motile forms in the surrounding fluid are extremely interesting and seeing them is not to be missed. You can expect to spend hours observing your infusion. In many ways this experiment is much more interesting than an almost instant preparation made of clay or a similar solid material, because we see the bions developing and the start of bionous movement as the grass breaks down.[7] It really is quite awe-inspiring to see the grass start moving and even more inspiring when a new organism breaks away from the grass and takes off on its own.

It will be much easier and safer to set up this experiment, if you have a helper at hand to hold items for you and, in particular, to take care of the melting waxes while you are concentrating on cutting up the grass blades. Remember that both these materials, paraffin wax and petroleum jelly, have quite low flash-points and if you absent-mindedly focus too much on the slide while the pan is getting hotter and hotter, you could have a dangerous fire on your hands. You can eliminate this danger, if your assistant looks after the pan for you as it heats up. (I often have to work without an assistant. I have solved the safety problem by heating my waxes up in an old Le Creuset cooking pan. This has a very heavy, solid base that retains a great deal of heat. I heat this up until it is as hot as it's safe to go and then switch the heat off and place the pan on a wooden mat. I can then get on with cutting up and gluing the grass with no fear of an accidental fire as there is no heat under the pan. The waxes stay molten for some time.) You also need to avoid any draughts. Once your precious strips of grass have been cut and trimmed they can easily be blown away, if someone cheerfully barges into your work-room, opening the door wide in a breeze.

Take a blade of grass, place it on your chopping board, and slice along the spine, the thick ridge of tissue running down the centre of the blade. Put the piece that you have just cut out carefully to one side and then slice along the other side of the spine. Discard the spine immediately, so you do not mix it up with the pieces that you want to use in the experiment. You should now have two pieces of grass about five centimetres/two inches in length and possibly a millimetre or two wide. You need to shorten these somewhat so that they will lie on the slide and protrude beyond the coverslip, so that you can anchor them to the slide surface with wax. See the illustration for details and measurements. (This drawing is more or less life-size.)

Before you do anything with your blades of grass, clean them carefully with an alcoholic swab, thus removing any 'spores' or contaminating bacilli. Trim your pieces of grass carefully to length and prepare to attach them to the slide. This step is the trickiest part of the job. I shall assume that your helper has got the waxes mixed and melted nicely. Bring the pan over to your work site and place it on an insulating piece of wood or similar, especially if you are working in improvised workspace, not in a lab. Have your paint-brush to hand as well. Place your first piece of grass in position, hold it there carefully, and put a dab of hot wax on its end so that it is glued to the glass. Repeat the process with the other piece, placing them both parallel to each other, so that their ends will almost reach to the edge of the coverslip when you position that. (See drawing, if this is not clear.)

Once you have got your strips of grass into position things get easier. Take a coverslip out of its container, grip it carefully with a pair of coverslip forceps and place it on the slide so that it covers the grass strips and leaves about a quarter of the slide-well open to the air. Again, the illustration should make this quite clear to you, if my verbal description is inadequate. You now need some tool that will hold the coverslip in position without breaking the delicate glass, while you paint round it with hot wax, thus securing it in position for the experiment. The perfect tool for this job is one of those school pencils with an eraser on the end. Hold the coverslip down carefully with the rubber while you paint the wax round the three sides of the coverslip. The wax should form a line about ⅛" wide, half of it over the slip and the other half on the slide surface. If you have got this far without losing your grass, breaking your coverslip, or making a mess on the slide with the wax, you are almost home and dry now. The last job to do is to paint a dyke of wax round the edges of the slide, so that any water that accidentally escapes from underneath the coverslip does not end up getting into the condenser or an objective. Remember that even on a cheap student microscope it may cost a few hundred pounds to replace a 100x objective and on more expensive instruments such an item may cost thousands. If you are working with a borrowed instrument… enough said.

Once your coverslip is safely attached and the dyke in position you need to make sure that there is no wax on the underneath of the slide. It must be able to sit flat and flush on the stage. Rub your thumbnail round the underneath edge or scrape it clean with your safety razor. Your slide is now ready for the experiment.

If you are going to be busy over the next few days, leave your actual experiment until you have got plenty of free time over two or three days for frequent observations. This experiment is so interesting and exciting, especially when you do it for the first time, that you do not want to miss important stages of developments on the slide. If you work a conventional five-day week, the best time to start your experiment by adding the water would probably be Friday evening.

Set up a syringe with a needle, draw up some cooled boiled water and add it to the slide drop by drop until the space under the coverslip round the grass-blades is full of water, ideally without any bubbles. You may have the odd obstinate bubble that you cannot get rid of. This usually disappears after a while. Once your slide has been started with the addition of water, you can keep it going almost indefinitely by simply adding small amounts of water to prevent drying out. Evaporation will obviously be much faster in hot weather than cold. Your first grass-infusion experiment is now started and under way. Record the date and time exactly.

Observations

This is a special moment in your orgonomic education and I do urge you to make sure you have got lots of time available for your observations. This is not just because the job demands extensive observing. Once you realise what is happening, believe me, you will find it difficult to tear yourself away from the microscope.

It is informative to take a look at the grass immediately after you have added the water, if not before, and get a clear impression of what the grass looks like before it has undergone bionous disintegration. Almost any natural structure is beautiful and interesting when magnified and anyway, you need to be able to compare the bionous surface with the normal appearance of the blades. Yet again, I suggest you draw the grass at this stage. You can see its structure best at a relatively low magnification. We will be observing the bions and bionous disintegration at high magnification, 1000x or 1500x/2000x, if available on your microscope, so draw the grass before we have got oil on the coverslip and have switched up to the high-power objective. One of the drawbacks to using oil-immersion is that you cannot remove the oil from the coverslip and go back to low power magnification. A great advantage to the magnification changer on C O R E's Olympus is that it gives us a magnification of x1600 without using oil-immersion. This proves wonderfully convenient and means that you can switch between this and lower powers if you lose an item on the slide and need to find it again. (This is much easier at low power.) You could make your observations for the first time on an SP100 or SP150 instrument using the 20x eye-pieces, the 40x objective, and the resulting 800x intermediate magnification, if you are wary about using oil immersion.

So…what do we see at this exciting moment? At 100x or 150x we can see the beautiful green tone of the intact grass cells and their regular rectangular structure. If you have just added the water within the last half an hour or so there will be next to no movement seen on the slide. After you have seen enough or drawn this structure, move up to intermediate magnification and take another look. You will see more of the same only in greater detail. Again, wander round the slide and look at any details that strike you. The more you notice, the better.

We already know how to set up high magnification with oil-immersion from our experience with the clay preparation. As long as we are careful not to get oil anywhere but in the centre of the coverslip, this is no particular difficulty. We place the drop of oil, wind the stage up again, see the little flash as the oil and objective lens meet, and then continue with the fine-focus wheel. The grass's structure comes into view in great detail as we turn the stage up. Find a point where the cell walls meet and take a good look at the structure. This is where we will see the first changes as bionous disintegration sets in and becomes visible. (See sketches, (centre pages),if you are not sure what I mean.)

A good time-table for these observations is to start your experiment one evening, which we will call day 1. This means that the time when the least is happening on your slide is your time in bed. The following day, day 2, will also be fairly quiet, though you should certainly be looking at it now and again during this day, in case anything unforeseen occurs. This plan means that you are awake from hours 36-48 after the start of your project. This is the time when there is the most to be seen, important if you are doing this for the first time. By 36 hours you should be seeing bionous areas appearing on the surface of the grass. These are recognisable by their grainy appearance. There are lots of dots on the grass surface and the longer the experiment goes on, the more of these you will see. You may also see some of these dots moving about independently in the surrounding fluid. Take a careful look at the edges of your grass blades. You may be able to see a swelling form bounded by a membrane at the edge full of these bions. (See sketch.) This is well illustrated in Reich's two books, *The Bion Experiments* and *The Cancer Biopathy*,[8] though many of the photographs are not very meaningful until you have seen the items in question under your own microscope. That was my experience anyway. If you look long enough and carefully enough, you may see one of these forms break away and move off independently. You will not want to miss this, so expect to spend a great deal of time observing on this day, day 3. What is going on may well keep you out of bed that night. I imagine most readers will read this book at least once before they get down to repeating these experiments and those who try them will not be many. You feel a rare excitement as you do them. However sympathetic you are to Reich's written work and however much you feel that all his ideas, observations, and theories are sensible and reasonable, you enter a peculiarly unique area of experience and emotion, when you carry out these projects and actually confirm Reich's findings, which, of course, you will. The bion experiments produce an emotionally powerful reaction. It is one thing to have read about the bions and to know the theory. Anyone can read a book, remember a scientific report, summarise the argument, and, if they wish, tell other people about it. But to repeat an experiment and to actually see with your own eyes the forms described by Reich and Brown is an experience of a different order altogether. I can't put a single word to this feeling: it is a mixture of vibrating curiosity, a sense of awe at seeing an unknown natural process for the first time, and the imaginative wonder one feels when looking at something in a museum that is exactly as it was a hundred, even two hundred years ago. We are seeing what Reich and his colleagues must have been observing in Oslo in the nineteen-thirties and what Brown observed and described in 1828. It is a strange feeling to be walking around knowing these things that contradict everything we are taught at school and in science books. It can be a difficult feeling to live with.

Although Reich is still ignored by orthodox science, enough orgonomists have repeated these experiments for us to be able to say that the findings that he first reported on in his books are quite solid and scientific.[9, 10] By accident, because orgonomists are so spread out and isolated, there is another layer of evidence in our work in that we have not been looking over each other's shoulders while we do this work. We, as far as I know, have all done it, to start with independently, on our own and written down our results before comparing notes with other workers. Also some of us have repeated some of the experiments publicly at conferences in the public eye in front of trained biologists and other scientists who have been able to question us, examine equipment, and generally criticise and interrogate as much as they wish. In my own case, and I think this is probably typical, when I have been writing about my findings and my memories are uncertain about something, I set up the experiment in question and repeat it yet again, so that I usually have what I am describing before my very eyes, as I write.

So… it is now day 2, in the morning, and we get up take a look at our culture which has been running for about 12 hours. We have already had a good look at it immediately after adding the water, so we know what the grass looks like at various magnifications. We have switched up to oil immersion so we can now only observe at high magnification. There does not seem to have been any major changes during the night, so we take our time and draw the surface of the grass at this high magnification. The drawing looks more or less like the sketch in the centre pages.

We look regularly during the day. We have a last look at about 9:00pm before we go to bed. All this close observing is hard work and tiring on the eyes. Ah! Here and there close to the points where the cell walls meet there seems to be a dot that was not there before. This may, of course, be a grain or two of dust that has got into the preparation from the air or just a grain of soil that was attached to the grass all along. Let's not get too excited. The slide is, after all, open to the air. We'll have another look in the morning, just to make sure. But it's exciting all the same and we wake up in the middle of the night wondering how the slide is doing. There is nothing for it. We'll just have to go downstairs and have a look at it. Wow! There are now four or five times as many dots as we saw before and you rub your eyes to make sure you are not dreaming. You feel sure that you have just seen one come into existence before your very eyes. No, it can't be real can it? You must be dreaming. Let's go back to bed and have a fresh look in the light of day. Then you remember about the edge of the grass. You move the slide carefully in one plane so that you can find those dots again. As you slowly move across the surface you see lots more of the dots, in similar places along the lines of the edges of the cells and then you come to the edge of the grass. You are now looking at a green strip, the grass and clear water. There are a few items dancing and swimming around in the clear water, but what's going on at the edge of the grass? It can't be…no! But it is! Exactly as you remember from the pictures, you can see little swellings full of other dots.[11] This is just too much. You sit down, freezing in your night-clothes, but you can't resist it. These items have an irresistible magnetism about them. Hell! Is this what a career in orgonomy is going to mean, getting up in the middle of the night all the time to observe things under microscopes? Yes, it may well mean that, unless you are fortunate enough to attract a team around yourself, so that people can take it in turns to record observations. You gawp down the microscope, transfixed. (I did warn you, didn't I?) You are now missing sleep and not even noticing it. You'd swear one of those bulges on the edge of the grass is moving. You can't be sure. The bions within the membrane are moving vigorously. Again, you decide to leave it till the morning, when you are feeling a bit brighter, a bit more awake. Eventually you straighten up, yawning, and realise that you have been observing for an hour or longer. And so…reluctantly, back to bed.

Next morning the slide has been running for 36 hours or so. The next 12 hours are going to be the most interesting time yet of your orgonomic studies, if not of your life. If you haven't arranged things so that you can have a look at least every hour you are going to be missing some very interesting events taking place on your slide. Hurry your breakfast, no on second thoughts look at the slide first: breakfast can wait. You can eat breakfast any time.

So…what do we see at this stage? We sit down enthusiastically, impatient to see developments. Everything is there, in position. We switch on, remove the covers and…the dots are now so many that they seem to be swarming almost. The very surface of the grass is moving up and down slightly. Pairs of bions are dancing energetically around one another as in the clay experiment. At the edge the swellings that we saw in the night are bulging more than ever and the internal movement is more vigorous.

It is a good idea at this point to get out the text of *The Bion Experiments*[12] and *The Cancer Biopathy*[13] and to have another look at the photographs in those texts and to read through what Reich writes there. If you move around your slide you will not have to go very far before you come across a form that appears to be floating above the grass but which nevertheless seems to be attached to it still. This will probably be twitching and jerking, as if it is trying to break free from the matrix, the grass itself. You may be looking at a Volvox, which remains attached to a substrate. Try to fix some points of reference so that you can find this item again when you come back to your slide. An amoeba-like organism may eventually break free from the edge of the grass matrix as an independent organism – an exciting process to see, to put it mildly. Notice my qualifiers here. We are dealing with living processes here and nothing is completely predictable and black and white. So few reliable and serious observers have done these experiments that you may well see something that no-one else has noticed before. For reasons as yet unknown, we also do get bion preparations that just do not work, that produce no bions, though I have not seen this happen with a grass infusion. It appears to be a fairly reliable and predictable project.

There is no formal end to this experiment. You can keep on adding water to the preparation for days, even weeks, though, obviously the first few days are the most interesting and exciting. I have kept grass infusion slides going for several weeks. On one occasion the wax dyke round the edge of the slide broke down, but this caused no major problems. I was able to repair it with some freshly heated wax mixture. One possible limiting factor is the level of contamination from air-borne bacteria. These will inevitably get into your slide and multiply and eventually obscure the grass and bionous processes, especially in warm weather.

Postscript 2010: While recently repeating this experiment with the intention of filming it, I found my grass-infusion slides were very quickly becoming contaminated by a massive growth of rod bacteria. The presence of these forms is nothing unusual and they are always the first air-borne forms that one sees in an open slide, but the growth on these occasions was much faster than usual. I made several attempts, thinking at first that that particular slide was the problem, but this massive early contamination repeated itself several times. I have therefore devised another experiment which is in fact even more conclusive demonstration of the fact that disintegrating sterile vegetable tissue produces motile, apparently living forms.

I prepared sealed vials containing infusions of chopped hay, boiled them 10 times with a few days between each boiling, autoclaved them, and left them for a week in an incubator at 30°C. See the results on YouTube at http://www.youtube.com/watch?v=cXickkE3aDU I have repeated this experiment several times to check my findings and have confirmed them with every infusion. In one way this experiment is even more conclusive, as we know it is sterile until we open the vials for examination. These cultures produce all the motile forms described by Brown in his 1828 paper.

The Bion Experiment to End All Bion Experiments

When I wrote the original draft of this book I had not even done this particular experiment myself. It is so extremely simple and so devoid of opportunities for evasion and explaining away the results, that it seems an important one for any student of orgonomy to try.

This last experiment is so simple to do that it could be at the start of this chapter. But it does need some familiarity with the microscope and, more importantly, some confidence with the tools that we use around the microscope. So here it is, Reich's iron-filing experiment, as described by him in the early pages of his great book, *The Cancer Biopathy.*[14] This experiment is so simple, even naïve, that it is hard to imagine that Reich's account is true and that it is going to come out right, when we repeat it according to the information given in the book. The 'subject' of the experiment is a single iron-filing! But… it does exactly what Reich says it does in his book. This experiment is so simple and pure that it poses really difficult questions for the carping 'Brownian motion' merchants and the carriers of 'contamination.' They just cannot cite either of these two get-out clauses this time. The developments are far too quick for that, the movements too varied.

The perfect tool to pick up a single iron-filing is a pair of fine curved forceps from nhbs.com (item 185509 in their on-line catalogue). These have fine, pointed blades and make it fairly easy to pick up only a single filing, though even with these the operation demands care. To eliminate air-born contamination, you need to flame your slide and coverslip. To do that you need something to hold your slide and coverslip as you pass them in and out of the flame. A pair of larger metal forceps are the obvious tools. I use a pair of surgical forceps, which, for some strange reason, used to be on sale in Preston market. These look like large scissors and allow us to hold hot things. You could improvise with a large pair of ordinary metal forceps, but these might lose their grip and drop a hot slide on your or somewhere where it could do damage. You need to heat the glass evenly and slowly, so that it doesn't crack because of uneven expansion. Once you have exposed your slide to the flame all round, place it on some safe, non-combustible surface such as a block of wood, (not metal, which, if cold, may cause your slide to crack, because of eneven cooling). Now prepare your filing. Pick up as small a pinch as you can manage of filings and hold them in the flame until they have become red-hot. Drop a few onto the slide.

Remember, we want a single filing on the slide. You almost certainly won't manage that your first time. If you have got a few filings on the glass, look at the filings with a 10x magnifying glass and gently push off all but one or two with something sharp like a scalpel or razor blade.

This fiddly operation shows how helpful it is to have a low-magnification stereo-microscope in your lab, however improvised it is. Wiping the surplus filings off the slide is fairly easy under such a microscope, but more difficult using a magnifying glass. With a stereo-microscope you can even see enough detail to pick up a single iron-filing with the fine forceps. You can put it down, pick it up again, move it, and put it down again. I have just tried it myself to make sure.[15]

Now your single filing is on the slide, add a few drops of sterile 0.1N potassium chloride (KCl) or water, if you have no KCl. If you have got too much fluid on the slide, soak up some with a tissue. When the amount of fluid is right, drop a coverslip over the filing and place it on the stage. You can now start observing the most unexpected developments. Look at the filing at low magnification but do your main observations at high magnification. Here 20x eyepieces make your work much easier. Your objective is not in contact with the coverslip and any slight angle of the coverslip from the horizontal causes no problems. If you use your 100x objective with oil-immersion, the iron particle may lift the coverslip off the slide a little, making it difficult to get a well-focussed image. It can be done, but it requires care and caution. I have just checked to see that it is possible on the SP150 using 10x eyepieces and oil immersion with a 100x objective.

To begin with you will probably see nothing moving in the fluid surrounding your iron-filing. After about 5 minutes you should see single bions dancing about here and there. Within 10 minutes there will probably be plenty more. Follow the edge of your iron particle round until you find a 'bay' and you should see a real sea of bions dancing, cavorting together in pairs, and quite complex agglomerations of bions in advanced stages of combination. If you want to know what to expect or to check your findings, once you have done this experiment, you can see C O R E's video of the experiment on YouTube – *Bions from an Iron Filing (after Wilhelm Reich).*

And that's it. It's very simple. But the results are incontestable and irrefutable. If you show this to an ordinary biologist, be prepared for a strong, and probably irrational, reaction. (C O R E can send you a small quantity of KCl. Please contact us at info@orgonomyuk.org.uk .)

Repeating Brown's Observations of His Active Molecules

I suggest that you try a few other materials and become quite familiar with the characteristic motion of bions before you try any of

Brown's experiments. If you are carrying out these experiments with a Brunel SP100 microscope you can obtain 20x eye-pieces for this instrument (and for the cheaper SP40). These eye-pieces give you an intermediate magnification without oil-immersion of 800x. This is high enough for you to see bionous motility and to make out the shapes and the mode of movement of the smaller particles on your slide. This fairly high intermediate magnification means that you do not have to use oil-immersion for your studies to start with and this saves a beginner a great deal of bother. I use this magnification a lot when I am doing routine investigations, repeating a familiar experiment, or just checking a finding. Obviously you will want to use oil-immersion and your high-level magnification eventually to learn as much about the bions and their ways of moving as you can before setting up the grass-infusion experiment, but for day-to-day bion work, observation at 800x is quite adequate.

If you are lucky enough to have access to a microscope with even higher magnification, in excess of x3000, you may just be able to see the internal blue orgone energy charge and the internal orgonotic pulsation in your bions. C O R E's Olympus is equipped with a Leitz 160x objective. The magnification changer with 15x eye-pieces gives a magnification of 4800x, with 20x eyepieces 6400x. As far as I know, these very high-magnification objectives are no longer made by anyone and you will have to search for a second-hand one, if you wish to advance to this high-level work with the bions. Thanks to an alert orgonomic colleague, we found this marvellous objective on ebay. Some older microscopes have a built-in magnification changer. C O R E has an East German Zeiss instrument with a built-in 1.6x changer which with 20x eye-pieces gives a maximum magnification of 3200x. We found this on ebay, though I have not come across any other such model there or elsewhere.

So…we are now getting familiar with our microscope and what we can expect to see in various types of preparation. Let us move on to Brown's experiments. He first noticed an unexplained form of motion when examining pollen in water. How do we go about examining pollen and what do we see when we do? Ideally we should be using the pollen of evening primrose, if we are repeating Brown's investigations exactly as he did them. However, I have no evening primrose pollen to hand at present and have not been able to find any, so for the time being, let's use some other sort of pollen that is more easily available. A simple option is to buy some hive-pollen from a health-food shop. This is an expensive option, but since at the time it was the only easily available supply of pollen I bought a jar for C O R E's collection of bion materials. You may be able to beg a teaspoonful of this pollen from a friend who has some in their kitchen cupboard. This pollen comes in little waxen balls about 1-2 millimetres in diameter and to start the basic pollen examination you need exactly one of these balls! Unfortunately the smallest jar of this hive pollen contains far more than we need, 125 grams in weight, and costs about £10. A jar will last a few lifetimes. (I have since discovered that this pollen does not store well and soon goes moldy, so I recommend cadging a small amount from a friend, if you can find someone who has it in their kitchen.)

Anyway, we have got our sample of hive pollen from a shop or friend, so let us play about with that and see what we find. Place your little ball of pollen on a flat slide and have a look at it under your lowest magnification using the 4x or 10x objective. The hive pollen is a dark lump showing few details except that at the edges we can see a few pollen sacs peeping out here and there. We can readily recognise them as pollen. In this dry state the ball is quite hard and resistant to pressure. There are no moving things to be seen on our slide in this state using low magnification What happens if we wet it? We draw up a few millilitres of boiled water in a small syringe and place a few drops on and around the lump of pollen. A cloud of yellowness immediately starts to spread in the surrounding water. We can see this with the naked eye. This is promising. The pollen ball is not after all so resistant. We take a look down the microscope. There is already a large number of pollen sacs in the water surrounding the darker mass of the ball. There seem to be three predominant types of pollen. By far the most common on the slide is a rectangular type with rounded corners, about three times as long as it is broad. Far fewer in number, but also visible, are a few that look like bicycle cog-wheels and a few larger ones with smooth edges that look more or less like a nut with the odd seam across their surface. The yellow stain is spreading through the water.

We can only safely look at an item on a slide without a coverslip at lower power. At higher magnifications the objective is so close to the object that there is a danger that we will get water and bits of pollen in the objective. So we gently remove the main body of the pollen ball with a pair of forceps. If you have not got a pair of forceps, use the tip of a knife-blade. It is about ten minutes since we added the water. We carefully place a coverslip over the water and pollen and drop it down onto the slide. We take a look at low magnification to orientate ourselves and focus on an area where there is plenty of pollen. We then swing in the 40x objective, which gives us our comfortable x800 magnification without the bother of oil-immersion. Can it be? No, surely not! The slide appears to be swarming with bions and bionous particles already. This is a surprise. Nevertheless, we must just look open-mindedly at what is before our eyes and observe as accurately as we possibly can. Expectations should not play too big a part in our work, especially at this stage. And if our possibly excited expectations are confirmed quickly and easily, then all the more should we repeat our observations carefully and make quite sure that we are not allowing hopes or expectations to influence what we see.

This finding is most unexpected. Most of the pollen sacs appear to be intact, so where are the bions coming from? Presumably hive pollen does get mixed up with a wide variety of trace materials as the bees collect it from flowers and fly back with it to the hive. When it rains presumably rainwater washes in minute amounts of dust from the surrounding soil. And wind will stir up quite a large amount of the local soil particles, some of which ends up in the flowers and on the bees. These particles could produce bions easily and probably do, but this level of bionous activity cannot surely stem from such tiny amounts of material. The outer layer of a pollen grain is made of exine, a sort of cellulose, and is said in the textbooks and dictionaries to be very resistant to decay. Pollen grains thousands of years old are found intact in peat-bogs. For all that we cannot deny our findings. We must just leave them as they are for the time being.

It occurs to me that maybe the water from the kettle is teeming with bions and, unknowingly, we are injecting a copious bion supply when we add water to the slide. There are a few rust marks on the element that could produce iron bions and the odd bit of soil-dust or tea-leaves or other domestic items could get into the kettle. So, just to be sure, I place a few drops of water from the same syringe on another slide, place a coverslip on it, and examine it under the same microscope. I can hardly see anything solid in it at all. There is the occasional immobile speck and after a lot of looking, scanning up and down and from side to side, I find a single motile bion. So that is one possible explanation rejected.

The strange thing about this investigation and Brown's original article is that 'pollen grains' are by microscopic standards large bodies, far too large to show 'Brownian motion.' Brown even describes them as larger particles in his article. (Converted to μm the length he cites is approximately 6.25 – 4.5 μm.) Surely Brown cannot have seen his pollen particles moving in that way? If we read his sentence carefully he seems to be referring to particles within his pollen grains:

> This plant was *Clarckia pulchella*, of which the grains of pollen, taken from antherae full grown, but before bursting, were filled with particles or granules of unusually large size, varying from nearly $^{1}/_{4000}$ to about $^{1}/_{3000}$ of an inch in length, and of a figure between cylindrical and oblong, perhaps slightly flattened, and having rounded and equal extremities.

Whatever Brown meant to say, it is clear that the slide under our microscope here, now, before us, the fluid under our coverslip, is teeming with bions and also motile bionous particles and agglomerations of bions. There is even a motile particle that looks like a long cylinder and this is rotating on its longitudinal axis. Brown describes a similar movement of some of the particles in his investigation.

> In a few instances the particle was seen to turn on its longer axis.

He goes on to describe what sounded to me very like bions when I first read his paper.

> Grains of pollen from the same plant taken from antherae immediately after bursting, contained similar subcyclindrical particles, in reduced numbers however, and mixed with other particles, at least as numerous, of much smaller size, apparently spherical, and in rapid oscillatory motion.

Brown can be forgiven for assuming that these particles were apparently spherical. If he had looked a little longer and more attentively he would have seen that these particles and some of the aforementioned larger ones were not quite symmetrical and that he could have noticed their apparent shape changing rapidly as they moved. In other words they were spinning. As we know from our bion experiments, spinning is a characteristic motion of bions. Some of the small spheres in our preparation are also dancing about each other in pairs. I have repeated this investigation with a pollen ball of hive-pollen many times and got exactly the same findings each time.

So, let's try some 'real' pollen next, pollen straight out of a flower. I look round at the various flower stalls in the local market and freesias seem to be the smallest and cheapest. Let's start with them.

I scrape the stamens of the flowers with the sharp edge of a scalpel from a biology dissecting set and the green card I am using to catch the pollen on is covered with a fine light dust. (Notice the improvised tool here. Please don't think you must go out and buy a scalpel to do this experiment. I simply cast round for the best tool I could think of. Any metal item with a sharp edge would serve, an ordinary domestic knife, the blade of a penknife, or a razor blade.) There we are. We now have a pollen sample that we know is only pollen. I scrape up a little clump of the pollen with the same scalpel and tap it off onto a slide. I have already got a 2ml syringe of boiled water ready. I have a look at the pollen at low magnification with no water on it and it looks like any other pollen that I have seen, small rounded ovals by the dozen. I push out a few drops of water onto the pollen and place a coverslip over it. I look at it under higher magnification, x200, and already things seem to be happening. Once you are used to observing bions you can sense movement in the whirling dots even when they only look like dots. This is surprising again. I switch up to x800 and the more detailed picture confirms my hunch. The slide is full of bions dancing and spinning. There are even a few clumps of bions twitching and bending. I wander round the slide observing. I cannot see any obvious explanation for the bions. Later events show that I must have been a bit slow on the uptake here. I do notice that one or two of the pollen sacs seem to have a patch of granular substance coming out of them or from under them and wonder if this has something to do with the high bions levels. It is not clear where these granular materials have come from but I can only assume that they have come from within some of the pollen sacs. It dawns on me that I must be seeing what Brown describes as,

> …particles…of much smaller size, apparently spherical, and in rapid oscillatory motion.

So, it seems certain that Brown must have been observing and describing bions without knowing it. Clearly he had no conceivable framework into which these items could be fitted and so searched for an acceptable explanation. Someone of his time and background could not have imagined a life-energy, a process of bio-energetic charging, and the process of bionous disintegration. But, as we shall see, he was not that far from an intuitively correct understanding of them. According to Brown's biographer, Mabberley, he made these investigations while trying to discover what the energy was behind the process of fertilisation.[16] It seems that intuitively he was very close to the primary biological process of bionous disintegration and origin of life, which Reich later observed and described.

And while I have been doing all this, one of my contacts tells me she has got some evening primrose pollen for me. We shall now be able to repeat exactly the same investigation as the one that Brown made, as described in his original paper. (*Clarkia pulchella* is evening primrose.)

In the meantime a happy accident confirms the origin of the bions and bionous particles as occurring inside the pollen sacs themselves. While doing all this work on pollen I have become quite interested in the whole process of pollination and flower structure. I have learnt from the *New Naturalist* volume, *The Natural History of Pollination*, that the whole structure of a flower is the result of its adaptation to a particular method of pollination, pollination by an insect or the wind or water.[17] I am interested to learn more about the structure of flowers and while I am out on a walk along the river Ribble here I pick a single blossom from what is left of the Himalayan Balsam, which grows by the million along the river banks. As I peel it apart and examine the parts, I discover under the microscope that the anthers are covered with pollen sacs. I can conduct the same experiment with these as I have done with the freesia pollen. I scrape off the sacs with the scalpel blade again and have a look at low magnification. They look very similar to the freesia pollen sacs. I drop some boiled water from my syringe onto the grains and cover them with a coverslip. As I slowly focus and take a look at the slide an amazing sight meets my eye. The slide is already busy with movement and I see the sacs here and there exploding and spewing out a huge river of granular matter, like larva pouring out of a volcano. The smaller of these particles are clearly bions and are cavorting about, spinning and dancing in pairs occasionally and even at this early stage forming little motile clusters that bend and twitch. For some unknown reason the process by which the bions come into being occurs much faster amongst these pollen grains than it does in the freesia pollen. But they look exactly the same as those seen on the freesia slide, even though there has been no visible rupture of the freesia pollen sacs.

While out on another walk I notice a solitary convolvulus flower left over from the summer and pick that to see whether it contains any pollen. The anthers inspected at low magnification appears to still have a good number of pollen sacs attached to them. I carry out my usual procedure with it, which has now become an experimental protocol. I add a few drops of water to it and drop a coverslip onto it. It is then ready for safe observation at a higher magnification. At 200x the slide looks more or less like the others that we have seen and described above. There are not, as far as I can see, any of the rupturing sacs casting granular material out into the surrounding water but there are already a few dancing dots. I switch up to x800 and inspect the edges of the anther and the pollen sacs in the surrounding fluid. Sure enough there are the usual larger spherical particles which Brown describes and which do seem to show Brownian motion, the true pollen grains from within the pollen sac, but in between them there are plenty of tiny dancing dots and these already show agglomerations that migrate independently and contract and bend.

And now to the evening primrose pollen. This is getting exciting. Here we are really treading in Brown's footsteps. My guess is that this pollen will behave in exactly the same way as the other three varieties that we have already sampled.

That is exactly what the sample of evening primrose pollen does. My flowers are rather bedraggled and dead. I manage to scrape off a small number of pollen particles that I can see onto the slide. So, enough to carry out Brown's investigation, at least. We add the usual few drops of boiled water, drop a coverslip onto it, and place it on the stage again. By sheer good fortune the 10x objective is looking straight at a little cluster of pollen grains as we rack up the stage and the pollen comes into focus. (We are still using the 20x eyepieces, so we are now seeing everything at 200x.) As we look a grain bursts and spews forth an avalanche of grainy matter into the surrounding water. As it spreads into the water tiny dots at the advancing edge separate and are already spinning and dancing, several are dancing round one another and there are already a few agglomerations twitching and bending. As if to prove Brown right, a long, narrow item can be seen to be turning on its long axis. Brownian motion? Yes! But not in the sense that modern textbooks define it. It is Brownian motion, as originally observed by Brown, as if his original understanding of it had been accepted in his day, now what we must call bionous motion or motility.

I had wanted to repeat exactly the same investigation as the one that Brown had carried out with evening primrose pollen to be certain that we had repeated his actions and, as far as is humanly possible, were looking at exactly the same forms as those which he describes in his original paper. He seems to have been a very accurate observer. His descriptions of his items under his microscope are immediately recognisable. Once we have confirmed his examination of the behaviour of evening primrose pollen we know that it behaves in more or less the same way as the other species whose pollen we have investigated. This means that the student of orgonomy who wishes to investigate Brownian motion and compare it with bionous motility can examine the pollen of any easily available flower.

After Brown had noticed this strange motion of the particles originating from pollen grains in water, he investigated pollen from dead plants to see if this still produced his active molecules. It did. He had access to museum specimens and was therefore able to find samples from plants that were up to a hundred years old. He moved on to grass and moss and eventually to many different types of solid ground materials to see if, as he thought, they contained, or were made of these same forms, what he calls active molecules. To the student of orgonomy who has repeated Reich's bion experiments Brown's findings are no surprise at all. Brown's pathway was remarkably similar to Reich's.

It seems that, unawares, I was doing almost exactly the same as Brown did, when I first started repeating Reich's bion experiments. Once I had repeated the basic experiments described by Reich, I realised that this was a natural process that must be occurring all the time almost everywhere and that it would be interesting to test different materials for bionous disintegration and the production of bions. I collected small samples of every material I could get my hands on and tested them by doing the simple bion experiment as described above (pages 79-88). Brown did the same. At the time of writing I have already tested many items specifically mentioned by Brown, notably ground glass and various rocks. I had just received a sample of fossils from the Charmouth fossil shop in Dorset and this is exactly the sort of material that Brown tried.

Fossilised materials need extra preparation as they are extremely hard and need to be broken down into fairly small particles before we can grind them into powder in a pestle and mortar. I have devised a primitive method for doing this. Perhaps some readers will think of a simpler method than mine. I would be pleased to hear of it, if they do.

If you try and break up a fossil (or any very hard material) with a hammer, it shatters and shoots all over the place. Clearly we need a method that offers us some control over the bits as they break. Compressing a small lump of fossil slowly in a G-cramp is much better and more effective, though the bits still tend to shoot in all directions as the original piece crumbles. If we do the crushing in a thick brown-paper envelope, the envelope catches the fragments and the dust created. Bit by bit we end up with a collection of small fossil fragments and quite a lot of fine fossil dust, even before we have started grinding in the mortar. Once you have got, say, a dessertspoonful of bits and dust, pour them out into your mortar. You can now start grinding, though even after all this effort you may still have to carefully crush some of the larger pieces between the pestle and mortar. It is advisable to protect your eyes while doing this, as tiny pieces of hard fossil may pop and jump as you grind. Once you have got a modicum of finer dust, you can filter the contents of your mortar through a small strainer and then set about grinding these sieved particles really finely. You don't need a large amount to conduct one experiment, though all this is so messy and such a chore that it is probably best to keep at it and produce a usable quantity for future experiments, not just the current one. If your orgonomic work develops you may well find yourself both repeating these investigations to check your findings and also demonstrating them to other interested students.

We have obtained a small pile of fossil powder and can now start our project. Once we have got this far, the experiment is almost the same as the one described above, the experiment with red clay or rock-dust. Have a similar array of items ready and heat a pinch of the powder on a spatula. Remember that even though this material is all mineral and so will not actually burn, there is always a chance that something very hot will jump and spit at you or just fall off your spatula. If you have no dedicated laboratory facilities, do this heating over a kitchen stove with safe, metal surfaces that cannot catch fire. Once you have got your powder up to red-heat, keep it there for a couple of minutes, let it cool a little and then drop it into your test-tube that you have already filled with previously boiled water. Give it a good shake and close the tube with a cotton-wool ball while it cools. While doing this part of your experiment, keep small children and animals out of harm's way. If you are working with improvised facilities, possibly a kitchen table, or even worse a dining table, please take sensible precautions against injury to yourself and damage to the furniture. (Have a fire extinguisher or a fire-blanket to hand, in case of accidents. If short of money, you can improvise an effective fire-blanket by soaking a tea-towel in water and squeezing it out till it is not dripping. If anything catches fire, simply throw it over the flames before they take hold.)

Now, if Brown's findings, as handed on in the biology textbooks, are correct, we should see all these little fossil particles vibrating with Brownian motion, that is all particles below a certain crucial size should be moving randomly in all planes as they are bombarded by the moving molecules of the water. What do we see?

At first sight, at 80x the slide looks fairly similar to the clay/rock-dust slide. We see obviously solid particles of something mineral and crystalline, of a dark grey colour. In between the larger particles we see many tiny dots, which appear to be moving, though this is difficult to make out definitely at this magnification. My experienced eye can sense movement and spinning, even at this low magnification, though obviously we cannot be sure at this level.

At intermediate magnification the scene becomes even more familiar, though there is not quite as much movement as there was in the clay preparation. But at 800x on our Brunel SP100 the bionous motion is already quite visible. We check on C O R E's Olympus at x800 and we can also see and recognise bionous motility. Let us move up to high magnification and have a closer look at these tiny motile particles.

Sure enough, we see single bions tumbling in all planes, some spinning vigorously, and the familiar pairs cavorting around each other. There are also agglomerations, twitching, jerking, and moving about independently. While I am observing, I notice a pair of bions joined by a radiating bridge and a third bion close by dancing round them. It gets closer and closer and suddenly veers into them and joins the agglomeration. This is the first time that I have ever actually seen this process occur, even though there are usually many such agglomerations in any bion culture. We can also make out immobile particles similar in size and appearance to the motile ones. According to the theory of Brownian motion, these particles should all be moving. If we have already made it our business to recognise Brownian motion (which you should have done by now), we know that the particles bob up and down towards us and away from us and also vibrate in every other direction. Particles subject to Brownian motion do not spin, tumble, dance, twitch and turn on themselves, or form agglomerations.

I take a look at this slide a couple of hours later and the bionous motility has almost completely disappeared. The explanation for this may be that the bions have drifted to the bottom of the film of water on the slide and stopped moving as they land on the glass. We often see this happening in bion preparations. I turn back the focus wheel, so that I am now looking at the top of the film of water. Earlier there had been lots of motile bions floating about at this level. There are still 'dots' and agglomerations and other particles floating on the top of the water here, but they are all quite immobile now. I find one single still-motile bion on the slide. Now, if the earlier motion was Brownian motion how do we account for the lack of movement now, even amongst particles that are clearly floating on the top of the layer of water? It is a common empirical finding in bion studies – that some materials produce bions with a short life that soon stop moving altogether. As far as I know, no orgonomist has found an explanation for this yet. Reich himself mentions this difference in the life-time of bions from different materials and says he has as yet no explanation for it, apart from suggesting the reasons as chemical composition and the culture medium.[18] He does not appear to have considered pH levels as important, as has been demonstrated by Palm and Döring.[19]

Repeating this investigation with a sample of ground fossilised wood produces similar results. We do it first without sterilising or heating anything. We repeat it again, but this time we heat the ground fossil dust to red heat on a spatula before we add it to the water. This sample produces a much higher level of bionous growth than the unheateed sample, as Reich also found. Brown does not appear to notice this difference. At least he does not mention it in the article contained in the appendix. (He devised the same control as Reich did a hundred and ten years later.)

This may be some freak result. We cannot dismiss a claim that has apparently stood the test of a century on the basis of a single experiment, can we? So…let us try another material cited by Brown in his report. How about ground glass? That sounds as unlikely as iron filings, doesn't it? Unlikely or not, if we are real scientists, we should go ahead and try.

Glass needs even more caution and care in grinding than fossil material. Start out with some fairly delicate, easily broken glass, such as an accidentally broken wine-glass or a cheap, light glass bottle. If you are starting with an intact glass item, a safe way of breaking it is to place it inside an empty fruit-juice carton, bend the open end over carefully so no glass can jump out, put it on the floor and give the cartoon a good few blows with a hammer, until you can tell that the item has broken up into several much smaller fragments. After reaching this stage, pour the fragments into a strong envelope, as we did with the fossil, and proceed as above until you have got some quite small glass particles for your mortar. Protect your eyes when you get to that stage and make sure no particles can get anywhere dangerous. If you have small children or animals about your home, make sure they are safely excluded from your temporary work environment and for good measure sweep the floor carefully after you have prepared your ground glass, just in case some sharp fragments have ended up on the floor. After this, carry on as before as with the clay or fossil powder. The larger glass particles are quite visible in the water and the other findings are exactly the same as with the fossil or clay powders. I know this sounds unlikely, if not impossible, but I am only reporting what I myself have found after repeating Brown's own investigation.

Just to be sure, I try a preparation of copper filings. Brown says in his article that he has tried many metals. Heating copper on a spatula appears to change it to copper oxide. We can see it turning black as the heat spreads through it. We still find bions on the slide when we observe the preparation, not lots of them, but there are definitely some present, including agglomerations that bend and jerk. I have just, while digging out these old samples to repeat the experiments, counted my bion collection. It consists of about 70 different materials, including organic items such as chopped leaves of various species, ground organic matter such as cooked potato or dried orange-skin, and any number of ground minerals, including sand and rock from various places I have visited on holiday or which friends and relatives have sent me from places that they have visited. If this description tempts you to try these experiments, don't worry if you have no accessible samples of the items mentioned by name. Have a go yourself with something new. Anything that you can grind up finely will do.

Brown lists all the different items that he has investigated to see if they 'contain' his active molecules. I put the word in inverted commas, because that is how he saw his findings. To him, the molecules must have been within the materials for him to see them when the materials were ground up. He had no inkling at all, as far as we know, that the molecules were the result of a process, as Reich discovered later. Incidentally, this is a good example of the value of accurate, painstaking observations in practical science. Even if you do not recognise or understand what you are seeing, it is important to document your findings accurately. Then someone coming after you with more knowledge can, we hope, interpret your findings correctly. We do know that Brown was an exceedingly thorough and meticulous observer and only ever published his findings when he was sure of them, even though in this case, he could not explain them.

Apparently Brown showed one of his slides to Darwin just before he left for his voyage on the *Beagle*, but declined to say what it was that he had shown him.[20] Darwin did not follow up the puzzle and just accepted the mystery. What a wonderful time to have been living and working in, when one was invited to dinner and was shown something through a microscope at the very edge of science, before one sat down to eat!

Brown also investigated previously living material such as grass and moss. In other words he conducted the same investigation as Reich in his grass-infusion experiment. His findings were the same: he saw multitudes of his active molecules. He did not apparently observe his preparations long enough to witness the agglomeration into organisms or the emergence of clusters of active molecules/bions from the grass. If you have got this far you will already able to repeat the other investigations reported on by Brown in his original paper and those referred to in his *Additional Remarks*. If you have not got access to materials he tested but have other ones available, try those instead. The principle of the experiments and the orgonomic interpretation is the same and equally relevant, whatever the material. If you try a new material, you may discover something new and make some new links for orgonomic research.

Helpful books on the use of the microscope

Ford B (1973); The Optical Microscope Manual, David and Charles, Newton Abbot.

Oldfield R (1994); Light Microscopy – an Illustrated Guide, Mosby-Yearbook Europe, London.

Lacey A J (ed) (1989); Light Microscopy in Biology, IRL Press, Oxford.

---

[1] BEOL.

[2] CB.

[3] Walker P M B (ed) (1995); Larousse Dictionary of Science and Technology, *red heat*, page 920, Larousse, Edinburgh.

[4] Illingworth V (ed) (1990); Penguin Dictionary of Physics, *Brownian motion*, page 47, Penguin Books, London.

[5] CB, page 42-44, The Radiating Bridge between Two Orgonotic Systems.

[6] BEOL, pages 31-39.

[7] ibid.

[8] ibid and CB, pages 48-60

[9] DeMeo (ed) (2002); Pulse of the Planet, No 5, pp 79-113, papers on bions and biogenesis by Bernard Grad, Maxwell Snyder, Dong Chuol Kong and Hyun

Won Kim, James DeMeo, Orgone Biophysical Research Lab, Ashland, Oregon.

[10] Davidson D (2007); The Culturing of Bions from Uncooked Foodstuffs. Poster Presentation, C O R E's Fiftieth Anniversary Conference, Chipping, Lancashire. Also on video on YouTube: http://www.youtube.com/watch?v=YLCXg1_lzQM

[11] CB, figures 41-42.

[12] BEOL, pages 31-39 and figures 12-25.

[13] CB, pages 48-60 and figures 34-42.

[14] CB, page 25.

[15] Woolnough L (2010); Understanding and Using the Stereomicroscope, Queckett Microscopical Club, London.

[16] Mabberley D (1985); Jupiter Botanicus, page 268, Cramer, Braunschweig, and British Museum (Natural History), London.

[17] Proctor M, Yeo P, and Lack A (1996); The Natural History of Pollination, chapter 14, Pollination through Geological Time, Collins New Naturalist, HarperCollins, London.

[18] BEOL; page 114.

[19] Palm M and Döring D (1997); Neue Untersuchungen su den Seesandbionen von Wilhelm Reich, pages 562-580 in *Nach Reich*, (eds Demeo J and Senf B), Zweitausendeins, Frankfurt.

[20] Darwin C (1881); autobiographical text, page 60, republished in 2002 in Charles Darwin, *Autobiographies*, Penguin Classics, Penguin, London.

Appendix 1 Brown's *Active Molecules*

The objection to Reich's claims when he did his pioneering work in the nineteen thirties was that of Brownian motion. Robert Brown (1773-1858) was a gifted Scottish botanist who played an important part in the collection and classification of specimens and the development of botany in the first half of the nineteenth century. He travelled to Australia on behalf of the famous Sir Joseph Banks and assisted in the collection of vast numbers of plants for Banks's London collection. He later became its curator. His name lives on in science as the first person to notice that very small particles suspended in a fluid appear to be moving of their own accord. He was using the single lens microscope of his day, (though compound instruments had by then been in use for some time), and appears to have been a meticulous observer. He first wondered whether these particles were alive or whether the movement was purely physical, some basic property of small particles in such an environment. As we can see, if we read his original work on the subject, he was very open minded, approached the questions with no prejudices and simply investigated to see what he could see. He controlled his findings (in the modern scientific sense of that word) rigorously and was not able to come up with an explanation for this motion. His main check on his findings was exactly the same as Reich's – he heated his solid materials to high temperatures and still found they produced these molecules, as did Reich. (He does not state exactly to what temperature he heated his materials. Reich heated his to red heat, that is to a temperature in the range of 500-1000°C.)

For a finding to be accepted, we do not necessarily need an explanation; rigorous and accurate observation is enough. An explanation was not available until Einstein's lifetime; he worked out the mathematics of Brownian motion and explained it as the inevitable motion of small particles that were being unevenly bombarded by moving molecules of the fluid that they were suspended in. If a particle is large enough, larger than about 0.5- 1.0ηm in diameter, the bombardment evens itself out and we see no Brownian motion. Only smaller particles show this motion, according to Einstein's findings.

It is an excellent exercise for any student of orgonomy, (and of microbiology, too), to make it his business to examine a wide series of suspensions until he is sure that he has been able to observe pure Brownian motion. If you do this, you will then be certain that you are not confusing Brownian motion with orgonotic/bionous motility. You will be able to refute the blanket dismissals of your orgonomic findings that are typically laughed out of court with the words 'Brownian motion.'

Any student of orgonomy who reads Brown's original text will be much better informed than most biology graduates, who, I am sure, never bother to read this valuable and interesting document. It makes impressive reading and raises an important question: have the advocates of Brownian motion been taking his name in vain, actually misusing his findings to dismiss something that he actually discovered and which was in fact bionous motility? The answer to this question must surely be a definite yes. As we can see from the quotation on page 37, Reich clearly thought that conventional biologists were doing this.

I have reproduced the text exactly as originally printed, including the contemporary use of capitals for many nouns. Passages of special relevance to orgonomy and bion study are printed in heavy type.

Robert Brown

A Brief Account of Microscopical Observations made in the Months of June, July, and August 1827, on the Particles contained in the Pollen of Plants; and on the general Existence of active Molecules in Organic and Inorganic Bodies.

The observations, of which it is my intention to give a summary in the following pages, have all been made with a simple microscope, and indeed with one and the same lens, the focal length of which is about $^1/_{32}$ of an inch.[1]

[1] This convex lens, which has been several years in my possession, I obtained from Mr Bancks, optician in the Strand. After I had made considerable progress in the inquiry, I explained the nature of my subject to Mr Dollond, who obligingly made for me a simple pocket microscope, having very delicate adjustment, and furnished with excellent lenses, two of which are of much higher power than that above mentioned. To these I have often had recourse, and with great advantage, in investigating several minute points. But to give greater consistency to my statements, and to bring the subject as much as possible within the reach of general observation, I continued to employ throughout the whole of the inquiry the same lens with which it was commenced.

The examination of the unimpregnated vegetable Ovulum, an account of which was published early in 1826,[2] led me to attend more minutely than I had before done to the structure of the pollen, and to enquire into its mode of action on the Pistillum in Phaenogamous plants. In the Essay referred to, it was shown that the apex of the nucleus of the Ovulum, the point which is universally the seat of the future Embryo, was very generally brought into contact with the terminations of the probable channels of fecundation; these being either the surface of the placenta, the extremity of the descending process of the style, or more rarely, a part of the surface of the umbilical cord. It also appeared, however, from some of the facts noticed in the same essay, that there were cases in which the particles that were contained in the grains of pollen could hardly be conveyed to that point of the ovulum through the vessels or cellular tissue of the ovarium; and the knowledge of these cases, as well as of the structure and economy of the antherae in Asclepiadeae, had led me to doubt the correctness of observations made by Stiles and Gleichen upwards of sixty years ago, as well as of some recent statements, respecting the mode of action of the pollen in the process of impregnation.

It was not until late in the autumn of 1826 that I could attend to this subject; and the season was too far advanced to enable me to pursue the investigation. Finding, however, in one of the few plants then examined, the figure of the particles contained in the grains of pollen clearly discernible, and that figure not spherical but oblong, I expected with some confidence, to meet with plants in other respects more favourable to the inquiry, in which these particles, from peculiarity of form, might be traced through their whole course: and thus, perhaps, the question determined whether they in any case reached the apex of the ovulum, or whether their direct action is limited to other parts of the female organ.

My inquiry on this point was commenced in June 1827, and the first plant examined proved in some respects remarkably well adapted to the object in view.

This plant was the *Clarckia pulchella*, of which the grains of pollen, taken from antherae full grown, but before bursting, were filled with particles or granules of unusually large size, varying from nearly 1/4000th to about 1/3000th of an inch in length, and of a figure between cylindrical and oblong, perhaps slightly flattened, and having rounded and equal

---

[2] In the Botanical Appendix to Captain King's Voyages to Australia, vol ii, p. 534, *et seq*, (*anté p.* 435).

extremities. **While examining the form of these particles in water, I observed many of them very evidently in motion; their motion consisting not only off a change of place in the fluid, manifested by alterations in their relative positions, but also not infrequently of a change of form in the particle itself; a contraction or curvature taking place repeatedly about the middle of one side, accompanied by a corresponding swelling or convexity on the opposite side of the particle. In a few instances the particle was seen to turn on its longer axis, These motions were such as to satisfy me, after frequently repeated observation, that they arose neither from currents in the fluid, nor from its gradual evaporation, but belonged to the particle itself.**

Grains of pollen taken of the same plant taken from antherae immediately after bursting, contained similar subcylindrical particles, in reduced numbers, however, and mixed with other particles, at least as numerous, of much smaller size, apparently spherical, and in rapid oscillatory motion.

These smaller particles, or Molecules as I shall term them, when first seen, I considered to be some of the subcylindrical particles swimming vertically in the fluid. But frequent and careful examination lessened my confidence in this supposition: and on continuing to observe them until the water had entirely evaporated, both the cylindrical particles and spherical molecules were found on the stage of the microscope.

In extending my observations to many other plants of the same natural family, namely *Onagrariae*, the same general form and similar motions of particles were ascertained to exist, especially in the various species of Oenothera, which I examined. I found also in their grains of pollen taken from the antherae immediately after bursting, a manifest reduction in the proportion of the cyclindrical or oblong particles and a corresponding increase in that of the molecules, in a less remarkable degree, however, than in Clarckia.

This appearance, or rather the great increase in the number of molecules, and the reduction in that of the cylindrical particles, before the grain of pollen could possibly have come into contact with the stigma, were perplexing circumstances in this stage of the inquiry, and certainly not favourable to the supposition of the cylindrical particles acting directly on the ovulum; an opinion which I was inclined to adopt when I first saw them in motion. These circumstances, however, induced me to multiply my observations, and I accordingly examined numerous species of many of the more important and remarkable families of the two great primary divisions of Phaenogamous plants.

In all these plants particles were found, which in the different families or genera, varied in form from oblong to spherical, having manifest motions similar to those already described: except that the change of form in the oval and oblong particles was generally less obvious than in Onagrariae, and in the spherical particle was in no degree observable.[1] In a great proportion of these plants I also remarked the same reduction of the larger particles, and a corresponding increase in the molecules after the bursting of the antherae: the molecule, of apparently uniform size and form, being then always present; and in some cases, indeed, no other particles were observed, either in this or in any earlier stage of the secreting organ.

In many plants belonging to several different families, but especially to Gramineae, the membrane of the grain of pollen is so transparent that the motion of the larger particles within the entire grain was entirely visible; and it was manifest also at the more transparent angles, and in some cases even in the body of the grain in Onagrariae.

In *Asclepiadiae*, strictly so called, the mass of pollen filling each cell of the anthera is in no stage separable into distinct grains; but within, its tesselated or cellular membrane is filled with spherical particles, commonly of two sizes. Both these kinds of particles when immersed in water are generally seen in vivid motion; but the apparent motions of the larger particle might in these cases perhaps be caused by the rapid oscillation of the more numerous molecules. The mass of pollen in this tribe of plants never bursts, but merely connects itself by a determinate point, which is not infrequently semitransparent, to a process of nearly similar consistence, derived from the gland of the corresponding angle of the stigma.

In *Periploceae*, and in a few *Apocineae*, the pollen, which in these plants is separable into compound grains filled with spherical moving particles, is applied to processes of the stigma, analogous to those of Asclepiadeae. A similar economy exists in *Orchideae*, in which the pollen masses are always, at least in the early stage, granular; the grains whether simple or compound, containing minute, nearly spherical particles, but the

[1] In *Lolium perenne*, however, which I have more recently examined, though the particle was oval and of smaller size than in Onagrariae, this change of form was at least as remarkable, consisting in an equal contraction in the middle of each side, so as to divide it into two nearly orbicular portions.

whole mass being, with very few exceptions, connected by a determinate point of its surface with the stigma, or a glandular process of that organ.

Having found motion in the particles of the pollen of all the living plants which I had examined, I was led next to inquire whether this property continued after the death of the plant, and for what length of time it was retained.

In plants, either dried or immersed in spirits for a few days only, the particles of pollen of both kinds were found in motion equally evident with that observed in the living plant; specimens of several plants, some of which had been dried and preserved in a herbarium for upwards of twenty years, and others not less than a century, still exhibited the molecules or smaller spherical particles in considerable numbers, and in evident motion, along with a few of the larger particles, whose motions were much less manifest, and in some cases not observable.[1]

In this stage of the investigation having found, as I believed, a peculiar character in the motions of the particles of pollen in water, it occurred to me to appeal to this peculiarity as a test in certain families of Cryptogamous plants, namely, Mosses, and the genus Equisetum, in which the existence of sexual organs had not been universally admitted. In the supposed stamina of both these families, namely, in the cyclindrical antherae or pollen of Mosses, and on the surface of the four spathulate bodies surrounding the naked ovulum, as it may be considered, of Equisetum, I found minute spherical particles, apparently of the same size with the molecule described in Onagrariae, and having equally vivid motion on immersion in water; **and this motion was still observable in specimens both of Moses and Equiseta, which had been dried upwards of one hundred years.**

---

[1] While this sheet was passing through the press I have examined the pollen of several flowers which have been immersed in weak spirit about eleven months, particularly of *Viola tricolor*, *Zizania aquatica*, and *Zea maya*; and in all these plants the peculiar particles of the pollen, which are oval or short oblong, though somewhat reduced in number, retain their form perfectly, and exhibit evident motion,, though I think not so vivid as those belonging in the living plant. In *Viola tricolor*, in which as well as other species of the same natural section of the genus, the pollen has a very remarkable form, the grain on immersion in nitric acid still discharged its contents by its four angles, though with less force than in the recent plant.

The very unexpected fact of seeming vitality retained by these minute particles so long after the death of the plant would not perhaps materially have lessened my confidence in the supposed peculiarity. But I at the same time observed, that on bruising the ovula or seeds of Equisetum, which at first happened accidentally, I so greatly increased the number of moving particles, that the sources of the added quantity could not be doubted. I found also that on bruising first the floral leaves of Mosses, and then all other parts of these plants, that I readily obtained similar particles, not in equal quantity indeed, but equally in motion. My supposed test of the male organ was therefore necessarily abandoned.

Reflecting on all the facts with which I had now become acquainted, I was disposed to believe that the minute spherical particles or Molecules of apparently uniform size, first seen in the advanced state of the pollen of Onagrariae, and most other Phaenomagous plants, - then in the antherae of Mosses and on the surface of the bodies regarded as the stamina of Equisetum, - and lastly in the bruised portions of other parts of the same plants, were in reality the supposed constituent or elementary Molecules of organic bodies, first so considered by Buffon and Needham, then by Wrisberg with greater precision, soon after and still more particularly by Müller, and, very recently, by Dr Milne Edwards, who has revived the doctrine and supported it with much interesting detail. I now therefore expected to find these molecules in all organic bodies; and accordingly on examining the various animal and vegetable tissues, whether living or dead, they were always found to exist; and merely by bruising these substances in water, I never failed to ascertain their apparent identity in size, form, and motion, with the smaller particles of the grains of pollen.

I examined also various products of organic bodies, particularly the gum resins, and substances of organic origin, **extending my enquiry even to pit-coal; and in all these bodies Molecules were found in abundance. I remark here also, as a caution to those who may hereafter engage in the same enquiry, that the dust or soot deposited on all bodies in such quantity, especially in London, is entirely composed of these molecules.**

One of the substances examined, was a specimen of fossil wood, found in Wiltshire oolite, in a state to burn with flame; and as I found these molecules abundantly, and in motion in this specimen, I supposed that their existence, though in smaller quantity, might be ascertained in mineralized vegetable remains. With this view a minute portion of silicified wood, which exhibited the structure of Coniferae, was bruised, and spherical particles, or molecules in all respects like those so frequently mentioned, were readily obtainable from it; in such quantity, however, that the whole substance of the petrifaction seemed to be formed of them. But hence I inferred that these molecules were not limited to organic bodies, nor even to their products.

To establish the correctness of the inference, and to ascertain to what extent the molecules existed in mineral bodies, became the next object of inquiry. The first substance examined was a minute fragment of window-glass, from which, when merely bruised on the stage of the microscope, I readily and copiously obtained molecules agreeing in size, form, and motion with those which I had already seen.

I then proceeded to examine, and with similar results, such minerals as I either had at hand or could readily obtain, including several of the simple earths and metals, with many of their combinations.

Rocks of all ages, including those in which organic remains have never been found, yielded the molecules in abundance. Their existence was ascertained in each of the constituent minerals of granite, a fragment of the Sphinx being one of the specimens examined.

To mention all the mineral substances in which I have found these molecules would be tedious; and I shall confine myself in this summary to an enumeration of a few of the most remarkable. These were both of aqueous and igneous origin, as travertine, stalactites, lava, obsidian, pumice, volcanic ashes, and meteorites from various localities.[1] Of metals I may mention manganese, nickel, plumbago, bismuth, antimony, and arsenic. In a word, in every mineral which I could reduce to a powder, sufficiently fine to be temporarily suspended in water, I found these molecules more or less copiously; and in some cases, more particularly in siliceous crystals, the whole body submitted to examination appeared to be composed of them.

In many of the substances examined, especially those of a fibrous structure, as asbestos, actinolite, tremolite, zeolite, and even steatite, along with the spherical molecules, other corpuscles were found, like short fibres somewhat moniliform, whose transverse diameter appeared not to exceed that of the molecule, **or of which they seemed to be primary combinations. These fibrils, when of such length as to be probably composed of not more than four or five molecules, and still more**

---

[1] I have since found the molecules in the sand-tubes, formed by lightning, from Drig in Cumbria.

evidently when formed of two three only, were generally in motion, at least as vivid as that of the simple molecule itself; and which from the fibril often changing its position in the fluid, and from its occasional bending, might be said to be somewhat vermicular.

In other bodies which did not exhibit these fibrils, oval particles of a size about equal to two molecules, and which were also conjectured to be primary combinations of these, were not infrequently met with, and in motion generally more vivid than that of the simple molecule; their motion consisting in turning usually on their longer axis, and then often appearing to be flattened. Such oval particles were found to be numerous and extremely active in arsenic.

As mineral bodies which had been fused contained the moving molecules as abundantly as those of alluvial deposits, I was desirous of ascertaining whether the mobility of the particles existing in organic bodies was in any degree affected by the application of intense heat to the containing substance. With this view small portions of wood, both living and dead, linen, paper, cotton, wood, silk, hair, and muscular fibres, were exposed to the flame of a candle or burnt in platina forceps, heated by the blowpipe; and in all these bodies so heated, quenched in water, and immediately submitted to examination, the molecules were found, and in as evident motion as those obtained from the same substance before burning.

In some of the vegetable bodies burned in this manner, in addition to the simple molecules, primary combinations of these were observed, consisting of fibrils having transverse contractions, corresponding in number, as I conjectured, with that of the molecules composing them; and those fibrils, when not consisting of a greater number than four or five molecules, exhibited motion resembling in kind and vivacity that of the mineral fibrils already described, while longer fibrils of the same apparent diameter were at rest.

The substance found to yield those active fibrils in the largest proportion and in the most vivid motion was the mucous coat interposed between the skin and muscles of the haddock, especially after coagulation by heat.

The fine powder produced on the under surface of the fronds of several Ferns, particularly of *Acrostichum calamelanos*, and the species nearly related to it, was found to be entirely composed of simple molecules and their fibre-like compounds, both of them being evidently in motion.

There are three points of great importance which I was anxious to ascertain respecting these molecules, namely, their form, whether they are of uniform size, and their absolute magnitude. I am not, however, entirely satisfied with what I have been able to determine on any of these points.

As to form, I have stated the molecule to be spherical, and this I have done with some confidence; the apparent exceptions which occurred admitting, as it seems to me, of being explained by supposing such particles to be compounds . This supposition is in some cases hardly reconcileable with their apparent size, and requires for its support the further admission that, in combination, the figure of the molecule may be altered. **In the particles formerly considered as primary combinations of molecules, a certain change of form must be allowed; and even the simple molecule itself has sometimes appeared to me when in motion to have been slightly modified in this respect.**

My manner of estimating the absolute magnitude and uniformity in size of the molecules, found in the various bodies submitted to examination, was by placing them on a micrometer, divided to five thousandths of an inch, the lines of which were every distinct; or more rarely on one divided to ten thousandths, with fainter lines, not readily visible without the application of plumbago, as employed by Dr Wollaston, but which in my subject was inadmissible.

The results so obtained can only be regarded as approximations, on which, perhaps, for obvious reasons, much reliance will not be placed. From the number and degree of my observations, however, I am upon the whole disposed to believe the simple molecule to be of uniform size, though as existing in various substances and examined in substances more or less favourable, it is necessary to state that its diameter appeared to vary from $^1/_{15,000}$th to $^1/_{20,000}$th of an inch.[1]

I shall not at present enter into additional details, nor shall I hazard any conjectures whatever respecting these molecules, which appeared to be of such general existence in inorganic as well as organic bodies; and it is only further necessary to mention the principle substances from which I had

---

[1] While this sheet was passing through the press, Mr Dolland, at my request, obligingly examined the supposed pollen of *Equisetum virgatum* with his compound achromatic microscope, having in its focus a glass divided into 10,000ths of an inch, upon which the object was placed; and although the greater number of particles or molecules seen were about 1-20,000th, yet the smallest did not exceed 1-30,000th of an inch.

not been able to obtain them. These are oil, resin, wax and sulphur, such of the metals as I could not reduce to that minute state of subdivision necessary for their separation, and finally, bodies soluble in water.

In returning to the subject which my investigation commenced, and which was indeed the only I originally had in view, I had still to examine into the probable mode of action of the larger or peculiar molecules of the pollen, which, though in many cases diminished in number before the grain could possibly have been attached to the stigma, and particularly in Clarckia, the plant first examined, were yet in many other plants found in less diminished proportion, and might in nearly all cases be supposed to exist in sufficient quantity to form the essential agents in the process of fecundation. I was now therefore to enquire, whether their action was confined to the external organ, or whether it were possible to follow them to the nucleus of the ovulum itself. My endeavours, however, to trace them through the tissue of the style in plants well suited for this investigation, both from the size and form of the particles, and the development of the female parts, particularly Onagrariae, was not attended with success; and neither in this nor any other tribe examined, have I ever been able to find them in any part of the female organ except the stigma. Even in those families in which I have supposed the ovulum to be naked, namely, Cycadeae and Coniferae, I am inclined to think that the direct action of these particles, or of the pollen containing them, is exerted rather on the orifice of the proper membrane than on the included nucleus; an opinion which is in part founded on the partial withering confined to one side of the orifice of that membrane in the larch, - an appearance which I have remarked for several years.

To observers not aware of the existence of the elementary active molecules, so easily separated by pressure from all vegetable tissues, and which are disengaged and become more or less manifest in the incipient decay of semitransparent parts, it would not be difficult to trace granules through the whole length of the style: and as these granules are not always visible in the early and entire state of the organ, they would naturally be supposed to be derived from the pollen, in those cases at least in which its contained particles are not remarkably different in size and form from the molecule.

It is necessary also to observe that in many, perhaps I might say in most plants, in addition to the molecules separable from the stigma and style before the application of the pollen, other granules of greater size are obtained by pressure, which in some cases closely resemble the particles of the pollen in the same plants, and in a few cases even exceed them in size: these particles may be considered as primary combinations of the molecules, analogous to those already noticed in mineral bodies and in various organic tissues.

From the account formerly given of Asclepiadiae, Periploceae, and Orchideae, and particularly from what was observed of Asclepiadeae, it is difficult to imagine, in this family at least, that there can be an actual transmission of particles from the mass of pollen, which does not burst, through the processes of the stigma; and even in these processes I have never been able to observe them, though they are in general sufficiently transparent to show the particles were they present. But if these be a correct statement of the structure of the sexual organs in Asclepiadeae, the question respecting this family would no longer be, whether the particles in the pollen were transmitted through the stigma and the style to the ovula, but rather whether even actual contact of these particles with the surface of the stigma were necessary to impregnation.

Finally it may be remarked that those cases already adverted to, in which the apex of the nucleus of the ovulum, the supposed point of impregnation, is never brought into contact with the probable channels of fecundation, are more unfavourable to the opinion of the transmission of the particles of the pollen to the ovulum, than to that which considers the direct action of these particles as confined to the external parts of the female organ.

The observations, of which I have now given a brief account, were made in the months of June, July, and August, 1827. Those relating merely to the form and motion of the peculiar particles of the pollen were stated, and several of the objects shown, during these months, to many of my friends, particularly to Messrs. Bauer and Bicheno, Dr. Bostock, Dr. Fitton, Mr. E. Foster, Dr. Henderson, Sir Everard Home, Captain Home, Dr. Horsfield, Mr. Koenig, M. Lagasca, Mr. Lindley, Dr. Maton, Mr. Menzies, Dr. Prout, M. Renouard, Dr. Roget, Mr. Stokes, and Dr. Wollaston; and the general existence of the active molecules in inorganic as well as organic bodies, their apparent indestructibility by heat, and several of the facts respecting the primary combinations of the molecules were communicated to Dr. Wollaston and Mr. Stokes in the last week of August.

None of these gentlemen are here appealed to for the correctness of any of the statements made; my sole object in citing them being to prove from the period and general extent of the communications, that my observations were made within the dates given in the title of the present summary.

The facts ascertained respecting the motion of the particles of pollen were never considered by me as wholly original; this motion having, as I knew, been obscurely seen by Needham, and distinctly by, Gleichen, who not only observed the motion of the particles in water after the bursting of the pollen, but in several cases marked their change of place within the entire grain. He has not, however, given any satisfactory account either of the forms or of the motions of these particles, and in some cases appears to have confounded them with the elementary molecules, whose existence he was not aware of.

Before I engaged in the inquiry of 1827, I was acquainted only with the abstract given by M Adolpe Brongniart himself, of a very elaborate and valuable memoir, entitled "*Recherches sur la Génération et le Développement de l'Embryo dans les Végétaux Phanérogames*," which he had then read before the Academy of Sciences of Paris, and has since published in the *Annales des Sciences Naturelles*.

Neither in the abstract referred to, nor in the body of the memoir which M. Brogniart has with great candour given in n its original state, are there any observations, appearing of importance even to the author himself, on the motion of form of the particles; and the attempt to trace these particles to the ovulum with so imperfect a knowledge of their distinguishing characteristics could hardly be expected to prove satisfactory. Late in the summer of 1827, however, M Brogniart having at his command a microscope constructed by Amici, the celebrated professor of Modena, he was enabled to ascertain many important facts on both these points, the result of which he has given in the notes annexed to his memoir. On the general accuracy of his observations on the motions, form, and the size of the granules, as he terms the particles, I place great reliance. But in attempting to trace these particles through their whole course, he has overlooked two points of the greatest importance in the investigation.

For, in the first place, he was evidently unacquainted with the fact that, the active spherical molecules generally exist in the grain of pollen along with its proper particles; nor does it appear from any part of his memoir that he was aware of the existence of molecules having spontaneous or inherent motion, and distinct from the peculiar particles of the pollen, though he has doubtless seen them, and in some cases, as it seems to me, described them as those particles.

Secondly, he has been satisfied with the external appearance of the parts in coming to his conclusion, that no particles capable of motion exist in the style or stigma before impregnation.

That both simple molecules and larger particles of different form, and equally capable of motion, do exist in these parts, before the application of the pollen to the stigma can possibly have taken place, in many of the plants submitted by him to examination, may easily be ascertained; particularly in *Antirrhinum majus*, of which he has given a figure in a more advanced state, representing these molecules or particles, which he supposes to have been derived from the grains of pollen, adhering to the stigma.

There are some other points respecting the grains of pollen and their contained particles in which I also differ from M. Brogniart, namely in his supposition that the particles are not formed in the grain itself, but in the cavity of the anthera; in his assertion respecting the presence of pores on the surface of the grain in its early state, through which the particles formed in the anthera pass into its cavity; and lastly, on the existence of a membrane forming the coat of his *boyau* or mass of cylindrical forms ejected from the grain of the pollen.

I reserve, however, my observations on these and several other topics connected with the subject of the present inquiry for the more detailed account, which it is my intention give.

July 30th, 1828.

ADDITIONAL REMARKS ON ACTIVE MOLECULES.

BY ROBERT BROWN, F.R.S.

About twelve months ago I printed an account of Microscopical Observations made in the summer of 1827, on the particles contained in the pollen of plants; and on the general Existence of active molecules in organic and Inorganic Bodies.

In the present supplement to that account my objects are, to explain and modify a few of its statements, to advert to some of the remarks that have been made, either on the correctness or originality of the observations, and to the causes that have been considered sufficient for the explanation of the phenomena.

In the first place, I have to notice an erroneous assertion of more than one writer, namely that I have stated the active molecules to be animated. This mistake has probably arisen from my having communicated the facts in the same order in which they occurred, accompanied by the views which presented themselves in the different stages of the investigation; and in one case, from my having adopted the language in referring to the opinion of another enquirer into the first branch of the subject.

Although I endeavoured strictly to confine myself to the statement of the facts observed, yet in speaking of the active Molecules, I have not been able in all cases, to avoid the introduction of hypothesis; for such is the supposition that the equally active articles of greater size, and frequently of very different form, are primary compounds of these Molecules, - a supposition which, though professedly conjectural, I regret having so much insisted on, especially as it may seem connected with the opinion of the absolute identity of the Molecules, from whatever source derived.

On this latter subject, the only two points that I endeavoured to ascertain were their size and figure: and although I was, upon the whole, inclined to think that in these respects the molecules were similar from whatever source obtained, yet the evidence then adduced in support of the supposition was far from satisfactory; and I may add, that I am still less satisfied now that such is the fact. But even had the uniformity of the Molecules in those two points been absolutely established, it did not necessarily follow, nor have I anywhere stated, as has been imputed to me, that they also agreed in all their other properties and functions.

I have remarked that certain substances, namely, sulphur, resin, and wax, did not yield active particles, which, however, proceeded merely from defective manipulation; for I have since readily obtained them from all these bodies: at the same time I ought to notice that their existence in sulphur was previously mentioned to me by my friend Mr Lister.

In prosecuting the inquiry subsequent to the publication of my Observations, I have chiefly employed the simple microscope mentioned in the Pamphlet as having been made for me by Mr Dolland, and of which the three lenses that I have generally used, are of a 40th, 60th, and 70th of an inch focus.

Many of the observations have been repeated and confirmed with other simple microscopes having lenses of similar powers, and also with the best achromatic compound microscopes, either in my own possession or belonging to my friends.

The result of the inquiry at present essentially agrees with that which may be collected from my printed account, and may be here briefly stated in the following terms; namely,

That extremely minute particles of solid matter, whether obtained from organic or inorganic substances, when suspended in pure water, or in some other aqueous fluids, exhibit motions for which I am unable to account, and which from their irregularity and seeming independence, resemble in a marked degree the less rapid motions of some of the simplest animalcules of infusions. That the smallest moving particles observed, and which I have termed Active Molecules, appear to be spherical, or nearly so, and to be between 1-20,000dth and 1-30,000dth of an inch in diameter; and that other particles of considerably greater and various size, and either of similar or of very different figure, also present analogous motions in like circumstances.

I have formerly stated my belief that these motions of the particles neither arose from currents in the fluid containing them, nor depended on that intestine motion which may be supposed to accompany its evaporation.

These causes of motion, however, either singly or combined with others, - as, the attractions and repulsions among the particles themselves, their unstable equilibrium in the fluid in which they are suspended, their hygrometrical or capillary action, and in some cases the disengagement of volatile matter, or of minute air bubbles, - have been considered by several writers as sufficiently accounting for the appearances. Some of the alleged causes here stated, with others I have considered it unnecessary to mention, are not likely to be overlooked or to deceive observers of any experience in microscopical researches; and the insufficiency of the most important of these enumerated may, I think, be satisfactorily shown by means of a very simple experiment.

This experiment consists in reducing the drop of water containing the particles to microscopical minuteness and prolonging its existence by immersing it in a transparent fluid of inferior gravity, with which it is not miscible, and in which evaporation is extremely slow. If to almond-oil, which is a fluid having these properties, a considerably smaller proportion of water, duly impregnated with particles, be added, and the two fluids shaken or triturated together, drops of water of various sizes, from 1-50th to 1-2000dth of an inch in diameter, will be immediately produced. Of these, the most minute necessarily contain but few particles, and some may be occasionally observed with one particle only. In this manner minute drops, which if exposed to the air would be dissipated in less than a minute, may be retained for more than an hour. But in all the drops thus formed and protected, the motion of the particles takes place with undiminished activity, while the principle causes assigned for that motion, namely evaporation, and their mutual attraction and repulsion, are either materially reduced or absolutely null.

It may be here remarked, that those currents from centre to circumference, at first hardly perceptible, then more obvious, and at last very rapid, which constantly exist in drops exposed to the air, and disturb or completely overcome the proper motion of the particles, are wholly prevented in drops of small size immersed in oil, - a fact which, however, is only apparent in those drops which are flattened, in consequence of being nearly or absolutely in contact with the stage of the microscope.

That the motion of the particles is not produced by any cause acting on the surface of the drop, may be proved by inversion of the equipment; for by mixing a very small proportion of oil with the water containing the particles, **microscopic drops of oil of extreme minuteness, some of them not exceeding in size the particles themselves, will be found on the surface of the drop of water, and nearly or altogether at rest; while the particles in the centre or towards the bottom of the drop continue to move with their usual degree of activity**.

By means of the contrivance now described for reducing the size and prolonging the existence of the drops containing the particles, which, simple as it is, did not till very lately occur to me, a greater command of the subject is obtained, sufficient perhaps to enable us to ascertain the real cause of the motions in question.

Of the few experiments which I have made since this manner of observing was adopted, some appear to me so curious, that I do not venture to state them until they are verified by frequent and careful repetition.

I shall conclude these supplementary remarks to my former observations, by noticing the degree in which I consider these observations to have been anticipated.

That molecular was sometimes confounded with animalcular motion by several of the earlier microscopical observers, appears extremely probable from various pages in the writings of Leeuwenhoek, as well as from a very interesting paper by Stephen Gray, published in the 19th volume of the Philosophical Transactions.

Needham also, and Buffon, with whom the hypothesis of organic particles originated, seems not to have unfrequently fallen into the same mistake. And I am inclined to believe that Spallanzani, notwithstanding one of his statements respecting them, has under the head of *Animaletti d'ultimo ordine* included the active Molecules as well as true Animalcules.

I may next mention that Gleichen, the discoverer of the motions of the Particles of Pollen, also observed similar motions in the particles of the ovulum of Zea Mays.

Wrisberg and Müller, who adopted in part Buffon's hypothesis, state the globules, of which they suppose all organic bodies formed, to be capable of motion; and Müller distinguishes these moving organic globules from real Animalcules, with which, he adds, they have been confounded by some very respectable observers.

In 1814 Dr James Drummond, of Belfast, published in the 7th volume of the Transactions of the Royal Society of Edinburgh, a very valuable Paper, entitled "On certain Appearances observed in the Dissection of the Eyes of Fishes."

In this essay, which I regret I was entirely unacquainted with when I printed the account of my Observations, the author gives an account of the very remarkable motions of the spicula which form the silvery part of the choroid coat of the eyes of fishes.

These spicula were examined with a simple microscope, and as opaque objects, a strong light being thrown upon the drop of water in which they were suspended. The appearances are minutely described, and very ingenious reasoning employed to show that, to account for the motions, the least improbable conjecture is to suppose the spicula animated.

As these bodies were seen by reflected and not by transmitted light, a very correct idea of their actual motions could hardly be obtained; and with the low magnifying powers necessarily employed with the instrument and in the manner described, the more minutely nearly spherical particles or active molecules which, when higher powers were used, I have always found in abundance along with the spicula, entirely escaped observation.

Dr Drummond's researches were strictly limited to the spicula of the eyes and scales of fishes; and as he does not appear to have suspected that particles having autologous motions might exist in other organized bodies, and far less in inorganic matter, I consider myself anticipated by this acute observer to the same extent as by Gleichen, and in a much less degree than by Müller, whose statements have been already alluded to.

All the observers now mentioned have confined themselves to the examination of the particles or organic bodies. In 1819, however, Mr Bywater, of Liverpool, published an account of Microscopical Observations, in which it is stated that not only organic tissues, but also inorganic substances, consist of what he terms animated or irritable particles.

A second edition of this Essay appeared in 1828, probably altered in some points, but it may be supposed agreeing essentially in its statements with the edition of 1819, which I have never seen, and of the existence of which I was ignorant when I published my pamphlet.

From the edition of 1828, which I have but lately met with, it appears that Mr Bywater employed a compound microscope of the construction called Culpepper's, that the object was examined in a bright sunshine, and the light from the mirror thrown so obliquely on the stage as to give a blue colour to the infusion.

The first experiment I here subjoin in his own words.

"A small portion of flour must be placed on a slip of glass and mixed with a drop of water, then instantly applied to the microscope; and if stirred, it will appear evidently filled with innumerable small linear bodies, writhing and twisting about with extreme activity."

Similar bodies, and equally in motion, were obtained from animal and vegetable tissues, from vegetable mould, from sandstone after being made red hot, from coal, ashes, and other inorganic bodies.

I believe that in thus stating the manner in which Mr Bywater's experiments were conducted, I have enabled microscopical observers to judge of the extent and kind of optical illusion to which he was liable, and of which he does not seem to be aware. I have only to add that it is not here a question of priority; for if his observations are to be depended upon, mine must be entirely set aside.

July 28th, 1829.

Appendix 2 Extracts from *The Bion Experiments on the Origin of Life*, *The Cancer Biopathy*, and *Ether, God and Devil* by Wilhelm Reich

The Bions (*Die Bione*) was first published in German in Oslo in 1938. It did not appear in English until 1979 when it was published by Farrar, Straus and Giroux of New York under the above title. Much of the material contained in *The Bions* appears in summarised form in *The Cancer Biopathy*. Any serious student of orgonomy should read both volumes.

The title of chapter 1 *The Tension-Charge Formula* shows us that Reich was already seeing a connection between the orgasm formula, which he had already formulated in the early thirties and his work on the bions and bionous disintegration.

> Since the orgasm is an elementary phenomenon of life, the formula expressing it should also be demonstrable in the most primitive biological functions, for instance the vital functions of protozoa. The basic assumption was, therefore, that the orgasm formula is identical with the life formula. (Page 19.)

> The unity of function of the total organism seemed decisive to me here, i.e. the amoeba lives on in the metazoan in the form of the contractile and expansive vegetative apparatus. (Page 21.)

> On the seventh day, both disintegration into vesicles and the structuring had reached a very advanced stage. Even at a magnification of 700x it was possible to see sharply defined protrusions at the boundary of the angular, irregular, brownish-coloured lumps of earth. These protrusions looked like *vesicularly structured tubes*, alternately expanding and contracting; bending movements were also observed. At a magnification of 1625x I saw a brown lump of earth with vesicular protrusions at various points round its edges. It was linked with another clump of earth by a vesicularly striated mass. *At the connection point the clump of earth was bending and stretching.* At first I thought that I was mistaken, but further careful observation left no doubt: *the clump of earth was moving as if it were jointed: it was stretching and contracting*. After

another seven days the process of disintegration into vesicles, the formation of the striated structure, and the breakdown of the edges of the crystals into vesicles had advanced considerably. The protrusions along the edge of the crystal were moving in three different ways: (1) rotating round their axis: (2) stretching and contracting: (3) bending.

I have called these new formations on the crystals "plasmoids". (Pages 44-45.)

The Cancer Biopathy was first published as Part II of *The Discovery of the Orgone in 1948*. It was out of print for many years and very difficult to come by until Farrar, Straus and Giroux re-published it in 1973. It was also published here by Vision Press in 1974. They published several of Reich's titles at that time. 'Reichian therapy' was then becoming quite popular and fashionable, especially in London, and it seemed that interest in orgonomy was really beginning to take off here. It was not. Interest remained confined to the psycho-therapeutic sphere alone and *The Cancer Biopathy* was remaindered. I can still not forget the rush of feeling, part delight and part bitter regret, when I came across a copy of this valuable book for the throw-away price of £1:95 in one of London's well-known alternative bookshops. What a wretched fate for such an important and profound book. The lack of interest in this text at that time and its invisibility here betray Reich's status in this country. I have not yet met anyone in this country who has read this book. It was not mentioned in the recent Radio 3 program on Reich.[1]

I have chosen passages that illustrate the crucial points of this work. I hope these extracts will inspire readers to buy and read, even study the whole of this great work. All the extracts come from the early part of the book, the chapters on Reich's work on the bions and the actual discovery of the orgone. The later sections of the book are devoted to Reich's experimental treatment of the cancer biopathy with the orgone accumulator and the establishment of a healthy energy economy in the new-born baby. (Reich was the first worker to point out the importance of bio-energetic contact between mother and baby for the baby's health.)

Before proceeding to an investigation of other properties of the energy vesicles, we must establish whether the blue vesicles develop exclusively from carbon or from other substances as well. If they were to be found exclusively in carbon, the fundamental question concerning the nature of biological energy in non-living matter would be easy to answer. But the problem is complex, because the more substances we examine and subject to swelling, the more the following conclusion is confirmed; *All matter heated to incandescence and made to swell consists of or disintegrates into blue-glimmering vesicles:*...(Reich's own italics, page 20.)

A few fundamental problems must be cleared up before we can draw any conclusions from our observations. The concept of "Brownian movement" has been invoked as an objection to the claim that specific *bio-energetic* forces account for the motility of bions. Physicists have known for a long time that the smallest colloidal particles are in motion, i.e. that they move in the field in various directions. These movements have even been calculated. They are attributed to collisions between the molecules in the solution and the larger colloidal particles.

This interpretation is purely *physical* and *mechanistic*. Nothing within it is consistent with the biological energy manifestations of pulsation. Can this interpretation be applied to the phenomena observed in bionous energy vesicles? An interpretation is valid only if it makes new phenomena comprehensible. It is invalid if it conflicts with the observations. And when it directly contradicts the observations and can be replaced by another interpretation that offers a more satisfying explanation of the phenomena, it is useless.

The mechanical Brownian movement is defended by the physicists as dogma. Insofar as it is directed against mystical interpretations of living phenomena this defense is justified. But experience shows just as clearly that the "molecular movement" interpretation is itself not without irrational motives. Otherwise, the physicist who sees all around him only Brownian movements of a purely physical nature would not so stubbornly refuse to consider a few facts that contradict his interpretation in certain instances. I do not believe that I will ever be able to convince these physicists, but I know that the blind alleys into which the purely mechanistic viewpoint leads will one day force science to face up to new facts and arguments. (Pages 27-28.)

The fundamental question of all biology concerns the origin of the inner impulses in the living organism. No-one doubts that the living is distinguished from the non-living by the internal origin of the motor impulses. The internal motor impulses can be ascribed only to

> an energy active within the organism. The question of the origin of this energy itself is answered by the bion experiment.
>
> *The biologically effective energy, within the organism, that generates the impulses originates from the same matter as the bion.* I introduced the term "orgone" for the energy observable in motile, bionous matter, deriving it from the words "organism" and "orgastic" Henceforth the expression "orgonotic" encompasses all energy phenomena and processes specifically pertaining to the energy governing living matter. Every living organism is a membranous structure containing a quantity of orgone energy in its body fluids; it constitutes an "orgonotic system." (Reich's italics, pages 29-30.)

> There is probably nothing more suitable for the study of the function of tension → charge → discharge → relaxation than protozoa. Their movements, their plasmatic currents, their expansion and contraction speak an entirely unequivocal language in terms of our formula of the life function. (Pages 57-58.)

> There was no doubt of the existence of an energy possessing extraordinarily high biological activity. It remained only to discover what its nature was and how it could be measured. One of my colleagues told an assistant at the Bohr Institute in Copenhagen about the SAPA bions. This person considered the notion of bions from sand as so fantastic that I decided not to expose my new radiation discovery to the danger of a biased investigation, prejudiced from the start by fundamental disbelief. (Page 87.)

Ether, God and Devil is barely-known amongst the wider public. It is very well-known amongst students of orgonomy. It was first published in book form in 1951 and was republished later by Farrar, Straus and Giroux in 1973 in one volume together with *Cosmic Superimposition*. It is an outstanding introduction to orgonomic functionalism and explores the emotional dynamics behind the two main strands of thought in western culture, mysticism and mechanism. Reich also focuses on the fact that the orgone has been systematically ignored and 'non-discovered' so persistently over the centuries and wonders why this is so. Brown's study of active molecules is probably the most blatant example there has been of science's bumping up against the existence, effects, and functions of the orgone, and suddenly steering away from them as fast as it could go.

The emotions are the specific function of the living protoplasm. Living nature, in contrast to the non-living, responds to stimuli with "movement," or "motion" = "emotion." It necessarily follows, from the functional identity of emotion and plasmatic movement, that even the most primitive flakes of protoplasm have sensations. The sensations can be understood directly from the responses to stimuli. These responses from plasmatic flakes do not differ in any way from those of highly developed organisms. There are no lines to be drawn here. (Page 54.)

It is not true that I was the first to observe orgone energy and, with it, the functional law that merges organic and inorganic nature into one. In the course of two millennia of human history, people time and again encountered phenomena of orgone energy, or they developed thought systems that approached the reality of cosmic orgone energy. That these insights could not break through should be blamed on the same human character traits that created religious prohibitions and destroyed any progress in the right direction. Basically the weapons of destruction were invariably mechanistic, pseudo-scientific counter-arguments or mystical obscurantism, except in cases of physical annihilation. (Page 78.)

It is the fear of autonomic organ sensations that blocks the capacity to observe orgone energy. (Page 86.)

Functional thinking does not tolerate any static conditions. For it, all natural processes are in motion, even in the case of rigidified structures and immobile forms. It is precisely this motility and uncertainty in his thinking, this constant flux, which places the observer in contact with the processes of nature. The term "in flux" or "flowing" is valid, without qualifications, for the sensory perceptions of the scientist observing nature. That which is alive does not know any static conditions unless it is subjected to immobilization due to armouring. Nature, too, "flows" in every single one of its diverse functions as well as in its totality. (Page 98.)

---

[1] BBC Radio 3 (2011); *Sunday Feature* on Wilhelm Reich, broadcast on Sunday, March 13th, 2011.

Books by Wilhelm Reich referred to in the text and references (FSG = Farrar, Straus and Giroux, New York):

The Bioelectrical Investigation of Sexuality and Anxiety (BISA), original publication 1937, published in English in 1982 by FSG, and still available.

The Bion Experiments on the Origin of Life (BEOL), original publication 1938 in German as *Die Bione*, published for the first time in English in 1979 by FSG and still available.

The Function of the Orgasm (FO), original publication 1942, reprinted since several times and still available. The only work by Reich in print in the UK, (Souvenir Press, 1983, reprinted, 2012).

The Cancer Biopathy (CB), original publication 1948, reprinted in 1973, (1974 in the UK by Vision Press), still available as US paperback published by FSG.

Character Analysis (CA), first published 1933, and reprinted in two further editions in 1945 and 1948, of which the last is the definitive one, published by FSG and still in print.

Ether, God and Devil (EG&D), original publication 1951, reprinted as one volume with **Cosmic Superimposition** (CS) in 1973 by FSG and still available.

Selected Writings (SW), original publication 1960, reprinted since by FSG and still available. (Also published in the UK by Vision Press 1961 and 1973, now out of print).

Genitality in the Theory and Therapy of Neurosis (GTTN), originally published in German as *Die Funktion des Orgasmus* in 1927, this was first published in English in 1980 by FSG. Still in print.

Passion of Youth (PY), published in 1988 by FSG. An Autobiography 1897-1922. Still in print.

Beyond Psychology (BP), a collection of Reich's letters and journal extracts, edited by Mary Higgins, published in 1994 by FSG, still in print.

In comparison to conditions in the nineteen-sixties, when I started my study of orgonomy, students are now in paradise. Virtually all Reich's important titles are still in print and there are enormous numbers available on line second-hand. (See, for example, www.abebooks.co.uk) A search under Reich's name will throw up more than two thousand titles, some of them very cheap paperbacks. Current editions of these books are mainly available as print-on-demand paperback titles and so may take some time to obtain, though most are available on-line at the time of writing. Many suppliers listed on ABE offer their titles as new and available immediately.

Suppliers and Equipment

This list is, needless to say, not exhaustive. All the information given here is given in good faith with the intention of helping those curious to experiment to find the equipment that they will need. Buyers must be responsible for their own decisions and take all the information here as provisional. Most manufacturers update their products now and again and I strongly recommend anyone thinking of buying any of the items here to contact the suppliers or makers for up-to-date brochures or to look at their websites to find out if any more suitable models are now on sale.C O R E has no financial connections whatever with any of the firms mentioned below and the information is as objective as I can make it.

Fortunately the microscope market, especially the range suitable for the amateur, does not change while you are thinking about it, as does the computer market, and many basic models and items of equipment have been the same for longish periods. The sundries needed for a light microscope have barely changed for a hundred years. I make it clear where I have tested something myself or where I am just quoting information. The big item in all this is the purchase of a light microscope, unless you are lucky enough to know someone who is willing to lend you one. This is not a very likely possibility. Most microscopes belong to institutional laboratories in schools, colleges, universities, or hospitals. The private individual who has got an unused instrument in their attic or box-room is rare, but it is worth asking round. The budding orgonomist, which is what you are, if you are undertaking any of these experiments, must be enterprising, even inspired, in the solutions that she finds to problems. I suspect that there are large numbers of unused microscopes about, so-called obsolete models that have been put away in store-rooms and forgotten about. An 'obsolete' model that is twenty, even thirty, years old and has been used carefully and looked after well is still a very good microscope and will be suitable for the work we are doing here. Many such models find their way onto ebay. If you have no openings of this sort, then you will probably be best advised to buy a new instrument.

New Microscopes

As far as I know, there is no microscope shop anywhere in this country, that is, a shop where you can walk in, look at various microscopes from different makers, and buy one. You have to seek out suppliers who specialise in microscopes and accessories and from whom you can order by post or on-line. Incidentally, if you are thinking of buying a microscope, it is important to buy the very best you can afford. Whatever you do, please, for the sake of your own sanity, do not buy one of those dreadful 'kits' seen in shops in attractive boxes before Christmas. These are almost always quite useless, even as a basic beginner's instrument, let alone as a piece of real scientific equipment. If you buy a better quality instrument you will never find yourself wringing your hands over your stupid expense, but if you buy a really cheap, inferior instrument you will certainly find yourself wringing your hands, if not weeping in despair, at the difficulties encountered while struggling with an inadequate instrument.

There are many good-quality microscopes for sale on ebay. C O R E has bought several of these for students in our bion workshops. Most of these instruments appear to be in perfectly satisfactory condition and prices are quite reasonable, sometimes downright bargains. The market price on ebay of a binocular laboratory microscope with simple brightfield illumination and no extras appears to be around £150-£250, unless the item in question is clearly a very high quality microscope. Even some better-quality models go for this sort of price, so you really can find great bargains on ebay, if you familiarise yourself with their system and scour their listings well before bidding. Higher-quality instruments, for example, Leitz or Olympus, sell for more, around £300-£500, depending on condition, age, and accessories. A model with some unusual refinement may cost as much as £1000. To have such high-quality optics at your disposal for a few hundred pounds is still a huge bargain. A friend of C O R E's in France who knew little about microscopes, a beginner, has bought an old Olympus on ebay and is now getting on with useful orgonomic work. It can be done.

Suppliers of new microscopes

Brunel Microscopes, Unit 2, Vincients Road, Bumpers Farm Industrial Estate, Chippenham, Wiltshire, SN14 6 NQ. 01249 462655
www.brunelmicroscopes.co.uk

Brunel are a very helpful and informal firm and I recommend them unreservedly. You can get almost everything you need for this project from them. They are the nearest thing to the old-fashioned suppliers who used to sell all the separate chemicals and items needed to make your own slides and to preserve specimens. They also sell an interesting selection of books for the amateur microscopist. They have a small teaching room in their premises and you can visit them by arrangement and look at their wares without obligation. They encourage you to test their models in the teaching room. There is a great deal of information on their site aimed at beginners and learners.

C O R E bought, tested, and used at our 2007 conference two of Brunel's cheaper models. As far as I can see, these are the cheapest new models adequate for the basic bion experiments, the SP40 and SP100, costing about £350 and £573 respectively. The SP100 provides a lot more scope for improvement, if you cannot afford certain items at your initial purchase. You can for about another £300 buy a phase contrast kit for this instrument. The SP40 does not have phase contrast at high magnification and that facility is extremely useful in orgonomic microscopy. It allows you to see much that is invisible under bright field illumination. (Just before publication we bought and thoroughly tested Brunel's next model, the SP150. This has a field iris in the light source and plan objectives, which greatly improve image quality. This is an outstanding instrument for its price and the one we would recommend any serious student with limited funds to buy. The basic model costs at the time of writing about £725. It allows full Köhler lumination.) Beginners buying new optical equipment never regret buying too good an instrument, whereas they often end up feeling frustrated and thwarted by cheap, inadequate equipment and very quickly wish to buy a better quality instrument. Remember the better the quality of your microscope, the easier your work will be – unfair but indisputably true. Apart from its excellent optics, the Olympus BX50 is easy to use, beautifully machined and solidly and reliably designed and engineered. Cheaper, inferior instruments often demand a lot of juggling and care to produce good images. The Brunel instruments are unbelievably good value and quite easy to use. A second-hand Olympus in good working order, if you can get your hands on one, would be an excellent instrument for a beginner, as would any of the old classic Leitz or Zeiss models.

C O R E has just bought a microscope, new but technically second-hand, an 'unwanted gift' on ebay, for the unbelievable price of £145 and I have done a bion experiment on it just to check that its optics were adequate for the job. They were. Buying on ebay, if you have never used a microscope before, is risky, but a sensible thing to do if you have a friend who can check the item description for you.

We now have enough microscopes to run group workshops on the bion experiments. We have also bought a couple that we shall pass on at cost price to students of orgonomy. If you are a young student of orgonomy wishing to start these experiments, please contact C O R E. We may have a suitable microscope for you. (www.orgonomyuk.org.uk and info@orgonomyuk.org.uk)

Second-Hand Microscope Supplier

John Millham, 82, Brasenose Road, Didcot, Oxfordshire, OX11 7BN. 01235 817157, johnamillham@aol.com Millham sells on ebay, too.

Philip Harris of Hyde Buildings, Hyde, Cheshire, SK14 4SH, 0845 120 4520, enquiries@philipharris.co.uk , is a well-known firm of educational suppliers of scientific materials and equipment. They supply items by post but have a minimum order charge of, (at present), £40.

Mineral samples

As you will have seen from reading Brown's paper, he tested a large number of ground minerals during his research. If you want to check his findings, you will need samples from geological suppliers, though you can start on the common materials in your own locality. You will find that ingenuity will lead you to many materials easily available in your local shops, for example various sands in pet shops, ground limestone from garden suppliers, and so on. Builder's skips are a good source, too. Clay is commonly found in many areas, and you can also obtain some from a potter, if you know one, as I did. You may eventually decide you want to try some less common materials only available from a geological supplier. They will almost certainly be your only convenient source of fossilised wood. Small samples from suppliers are quite cheap, though of course, if they have to send them to you they are heavy, and so presumably delivery charges may be high. C O R E has bought a large range of samples from:

Northern Geological Supplies Ltd, 66, Gas Street, Bolton, Lancashire, BL1 4TG. 01204 389863, www.geologysuperstore.com . A visit to this firm is an enthralling experience, well worth the time and expense. When I visited them with a huge list of items collected from Brown's 1828 article, they had them all, even the sand-tubes formed by lightning. If you are really stuck, C O R E can send you small samples.

Information about the rock-dust sold for the organic revitalisation of soils can be obtained from www.seer.org.uk/purchase.htm The smallest quantity for sale is a 340g bag. If that frightens you, C O R E can send you a small amount, finely ground, suitable for a bion experiment. This material seems to be the best bion-producer for beginners to use. It produces bions reliably and generously, which may be why it does the soil so much good. (The traditional use of volcanic rock as a fertiliser by Italian peasant farmers is mentioned by Reich in *Beyond Psychology*, page 107.)

If you are going as far as setting up a real laboratory for yourself, the ideal home for your microscope is a bench of solid construction that does not vibrate or shake in use. Such items are expensive. You could, of course, improvise with a workshop bench or strengthen a solid domestic table with extra bracing between the legs and work-surface. Vibrations are grossly magnified under a microscope and make accurate observation difficult. This is particularly important, if your home is affected by heavy traffic or has flexible floors. (Floors built with wooden joists and planks move quite a lot when someone walks on them.) Even the tiny movements of your elbows on a fairly solid dining table are visibly magnified by your microscope at 1000x.

The minimum space that you need for this work is about two square foot on a table-top to accommodate a microscope and a small area round the microscope for your slides, coverslips, forceps, etc. I have tried this myself with one of C O R E's Brunel microscopes. If you need to move your equipment so that others can use the table, the microscope can be lifted and taken away without difficulty. (See illustrations.) You can put your sundries in a cardboard box so that you can clear them away at a moment's notice. You could also accommodate your equipment on the top of a bedside table or in the left-over space on a computer desk. It is obviously much easier to have a dedicated bench and work-space for your orgonomic microscopy, but you can still manage in an improvised laboratory. (A *Workmate* folding bench, from DIY shops, makes a good improvised bench for a microscope, though the legs are a bit awkward. Designed for use as a sawing horse, it is extremely solid and stable.)

Slingsby sell a small folding bench to provide a rigid work surface on buildings sites. This, when erected properly on a good surface, is unbelievably solid and would suit someone with limited space. The work surface measures 70cm x 100cm. The only drawback to this bench is its weight, (28 kilos), too heavy for a young person or even a smaller adult to set up or move single-handedly. If you have a regularly available helper or are not going to have to dismantle the bench very frequently, this would be a good lab bench for a beginner. C O R E owns and uses one. (Slingsby catalogue no – 331021DF6.) If you have got enough space for a small laboratory, Slingsby also produce an excellent work bench that can be dismantled easily. Setting it up or taking it down involves doing up/undoing about 16 nuts and bolts. C O R E bought one of these for use at our events and have found it makes an excellent solid microscope bench. The smallest size available is 75cm x 150cm (355147DF6). Our model is 70cm x 200cm (355149DF6). All the components are fairly light except for the worktop, which needs two people to handle it safely. These benches are not cheap, but they will last for years, appear to be quite indestructible, and provide really comfortable and safe working conditions. (www.slingsby.com)

The thing to do is to get a microscope and to get down to work. You can refine and improve your work conditions as you go along.

Orgonomy in the World Today and Further Study

In the hope that readers may wish to study orgonomy further, here is some information on orgonomy today. Do not expect glossy prospectuses from colleges running courses and handing out diplomas and degrees. Although far from dead, orgonomy at present is in a precarious condition. There is a small number of bodies teaching or conducting orgonomic research. Most are in the United States. In some countries there are hard-working, committed individuals who may be the only person in the whole country interested in orgonomy. People interested only in 'therapy' tend to call themselves 'Reichians' and to ignore his scientific work.

As far as I know, only two institutions conduct continuing training courses, The American College of Orgonomy. (www.orgonomy.org/) and The Institute for Orgonomic Science (www.ios). The first is a rather conservative organisation that accepts only trained psychiatrists for training as medical orgone therapists. The IOS has started to accept non-medical students for training who have a master's degree in a relevant subject. There is no formal training in orgonomy available in Britain. This is mainly because there is no interest in the subject. I have been trying to get courses established for decades without success.

This may discourage students with no degree and no chance of acquiring one or who are too old to start such studies. You can still study orgonomy on your own. All Reich's important books are available new or second-hand. Would-be students can contact me for help with the private study of orgonomy at info@orgonomyuk.org.uk. (See also C O R E's booklets, *A Student's Guide to Orgonomic Literature and Resources* and *Teaching Yourself Orgonomy*.) Almost any background is relevant to the study of the life energy. It is most relevant to medicine, science and areas such as psychology, psychiatry, biology, natural history, midwifery, child-development, infant care, education, cosmology, and music.

As with microscope work, my advice is to start and see how things grow of their own accord. This approach, strange to those brought up in western educational establishments, is very productive, as long as you notice opportunities available to you as you read. Just get started and one thing will lead to another. I started like that more than forty years ago and am still going strong. Students of orgonomy in your own country will give you advice and assistance. Above all, attend, if you can, some of the regular orgonomic conferences and make some orgonomic friends, people who you have met face to face. This is inspiring and makes up for the inevitable isolation in places such as the UK, Australia, and Africa, where you will probably be living a very long way from your nearest orgonomic colleague.

I have not included many addresses in this brief summary, as they can, anyway, be found easily on the internet. And finally a warning - expect a refusal to listen or look, expect eyes turned away in incomprehension, when you tell people about your interest. But the light in the eyes of the few who respond will make your work worthwhile.

Glossary of Orgonomic and Scientific Terms

Items marked * are specifically orgonomic expressions, though they may be used in a different sense in everyday English. Items marked + within an entry also have their own entry.

Active molecules: Brown's own name for the motile particles that he saw suspended in water. In Brown's day the use of the word *molecule*+ was much vaguer than it is now in science. It was used to describe any microscopic particle. See below for the present-day exact use of the word.

Amoeba: a common water-born protozoon that looks like a blob of gel with a few dots in it at low magnification. This is the creature that many school-children have shown to them or are made to draw by science teachers.

Anther: the male reproductive organ on a flower. It carries the pollen sacs,+ which contain the pollen grains.+ These were the items that Brown first observed in the experiments in which he noticed his *active molecules*.+

Armouring*: see muscular armouring*

Autoclavation: the sterilising of specimens or equipment by heat in a pressurised container, usually for thirty minutes at 120°C. The same result can be produced in a domestic pressure cooker.

Autoclaving tape: paper tape with light-brown diagonal stripes across it. It is used to seal packs of equipment before autoclavation. When the correct temperature and time have been reached the stripes turn a darker brown, so confirming true sterilisation. A useful item when autoclaving bion cultures. Stick a small strip of it on a vial or inside the lid of your pressure-cooker.

Autonomic nervous system (ANS): the nervous system that governs the many automatic, unconscious responses that the body produces as it meets various challenges or changes in the environment, both emotional and physical. For example, in anxiety our pulse rate increases markedly. This happens automatically: we do not need to make it happen. In a state of pleasurable relaxation the heart-rate and blood pressure fall. The first response is a *sympathetic*+ effect, the second a *parasympathetic*+ effect.

Bastian H C (1837-1915**)**: British physician and researcher, committed to a belief in spontaneous generation+ and heterogenesis.+ Author of many books, outmanoeuvred in the scientific infighting between himself and the opposition camp (Darwin, T H Huxley,+ Pasteur, et al), he was sidelined and ignored by the end of his life after being an influential and recognised worker in his field. He was a professor at University College London and had a great reputation as a teacher and pathologist. His main works in our field are *The Beginnings of Life* (1872) and *Studies in Heterogenesis* (1903).

Bio-energetic*: connected with, driven by, charged by, bio-energy.+

Bio-energy*: Reich's word for the life energy that he presumed existed before he had discovered the orgone+ in the late nineteen thirties. The term is used widely in the later chapters of FO.

Bion*: a very small, orgonotically charged energy vesicle,+ originating in the process by which solid matter swells in a fluid and undergoes bionous disintegration.+ (See BEOL.).

Bionous disintegration*: the process first observed and described by Reich in which matter swells and breaks down into highly charged, motile energy-vesicles.+ (See BEOL.)

Bionous motility*: the characteristic movement of these vesicles and also the movement of larger particles undergoing bionous disintegration.+ (See BEOL.)

Biopathy*: a term from orgonomic medicine and coined by Reich himself as the title of volume II of *The Discovery of the Orgone*: *The Cancer Biopathy*. A biopathy is a pathological condition of the energy economy of an individual and Reich's research demonstrated that the typical cancer patient has a particular energy pattern and character traits, noticeably a very low bio-energy level and a strong tendency towards resignation in the emotional realm. (See CB.)

Bright field: the basic mode of illumination in a biological microscope, in which light is directed upwards through the object on the stage without being modified in any way. We see the image against a background of bright light. It is the simplest form of illumination and is often the only sort available on cheaper microscopes.

Brownian motion: the movement imparted to smaller particles of matter suspended in a fluid by the kinetic energy+ of the molecules of the fluid. It only affects particles below a certain size of about 1 µm. The random motion of the medium's molecules cancel each other out above this size. It was not explained in physics until Einstein+ worked out the mathematics of such motion.

Buffon, Georges Louis LeClerc, Comte de (1707-1788): French naturalist and author of a huge 44-volume natural history. He collaborated with Needham+ and had a theory of *organic molecules*+ as the basic elements of life whose organisation in development was governed by the organism's *moule interieur* (internal mould).

Common functioning principle* (CFP): an orgonomic principle shared by two processes in nature which in conventional mechanistic biology might not be seen to have anything in common at all. (See EGD, chapter III.)

Cocci: a genus of micro-organisms, literally *spheres*.

Dark field illumination: a method of lighting in a microscope in which the light is directed across the line of sight rather than directly upwards and into the objective of the microscope. If there is an item on a slide the light is reflected upwards into the objective and the object appears to the viewer silver in a dark field, similar to a photographic negative. Such an image is very beautiful and the method is very effective at showing up very small items and movement.

Einstein, Albert, (1879-1955): German-born physicist, who emigrated to the USA in 1933 and worked there for the rest of his life, world-famous for formulating the theory of relativity. Awarded the 1921 Nobel Prize for physics. He is generally considered to be the greatest scientific genius of the twentieth century. He was the senior author with Leopold Infeld of *TheEvolution of Physics*. Their treatment of Brown's 1828 paper is discussed in chapter 6. (See Isaacson W (2008); *Einstein: His Life and Universe*, Pocket Books, London.)

Heterogenesis: the doctrine that dead but previously alive matter can break down and develop into living matter of a different species. The word is hardly heard nowadays, but was commonly used in the nineteenth century while the debate over spontaneous generation was still active. Two protagonists of the doctrine, Pouchet+ and Bastian,+ published books with the word in the title – respectively Hétérogénie (1859) and *Studies in Heterogenesis* (1903).

Huxley T H, (1825-1895): scientist, campaigner on behalf of Darwinian evolution, known as 'Darwin's bulldog' for his aggressive willingness to publicly engage with opponents. He also campaigned to improve the status of science and for improved government funding of science and scientific institutions. Not famous as a scientist, but a very influential figure in the development of science in the late nineteenth century. An active opponent of H C Bastian+ in the debate over spontaneous generation.+

Infeld, Leopold, (1898-1968**)**: Polish theoretical physicist, co-author with Einstein+ of *The Evolution of Physics*. Their treatment of Brown's 1828 paper is discussed in chapter 6. For more information about him see Wikipedia - http:en.wikipedia.org/wiki/Leopold_Infeld and his autobiography - *Quest: an Autobiography* (1968, reprinted 2006).

Kinetic energy: the energy possessed by a body by virtue of its motion.

Köhler illumination (often Koehler in English texts): a method of lighting that orgonomic microscopy should ideally have facility for. It requires a field iris in the light source in the microscope base. It sounds very technical to describe verbally, but is quite simple to produce. It is best explained in an animated video, of which there are several good examples on YouTube and on Brunel's website.

Leeuwenhoek, Anton van, (1632-1723**)**: often wrongly thought to be the inventor of the microscope, pioneering microscopist and maker of single-lens microscopes. Leeuwenhoek single-handedly established the sciences of microscopy and microbiology. He was the first to see the microscopical items whose existence we now take for granted – pollen,+ micro-organisms, water-borne micro-life, bacteria, sperm, blood cells, and so on. He was a prodigiously accurate observer and also a genius at designing and making microscopes. His work stood on its own for decades after his death, as his talents were so above those of any successors. His letters reporting his findings were published in *Philosophical Transactions*.+ (See *Antony van Leeuwenhoek and his "Little Animals"* by C Dobell (1960); *Single Lens* (1985) and *The Leeuwenhoek Legacy* (1991) by Brian J Ford.)

Membrane: Biology usually refers to cell membranes, the membrane surrounding any cell. I occasionally use the word loosely to refer to the 'outside' wall of a bion.+ As far as I know, no-one has studied this 'membrane' in detail. It would be very interesting and important to find out exactly what the apparent membrane surrounding a bion consists of. I imagine it will be a difficult item to investigate. Conventional biology would start doing this with the electron microscope, but the treatment needed to examine a bion in this way would destroy it. A common-sense assumption would be that the outside wall of a bion is a sphere of molecules of the material forming the bion – iron, mineral, or vegetable tissue, possibly held together by orgonotic attraction, the same force as gravity, according to Reich's gravitational theories. Some bionous materials, eg, soils and rocks, are complex mixtures which do not form discrete chemical molecules. So this hypothesis does not cover all possibilities.

Molecule: in Brown's day this word was used to describe any microscopically small particle. In modern science it has a precise meaning: the smallest quantity of a compound or element that can exist on its own. Thus an element, oxygen, has a molecule containing two atoms of the gas (O_2) and a molecule of a compound, for example sulphuric acid, contains two atoms of hydrogen, one of sulphur and four of oxygen, (H_2SO_4).

Muscular armouring*: permanent muscular tension that blocks both the free movement of orgone energy+ within the organism and the feelings that would be experienced if it were moving freely. This armouring comes about in infancy when primary emotional needs are not satisfied and the child tenses up to prevent the very deep pain resulting from this frustration.
Needham J T, (1713-1781**)**: English Catholic priest who worked mainly in France and Belgium and published his work in French, though papers by him also appeared in English in the *Philosophical Transactions*.+ He collaborated with the Count de Buffon+ and claimed to have demonstrated the origin of living forms in sterile preparations.

Orgasm formula*: the cycle of *mechanical tension → bio-energetic charge → bio-energetic discharge → mechanical relaxation* occurring in the sexual embrace and which Reich later realised was a basic function throughout animal nature in organs and systems. He then named it *orgonotic pulsation*. (See the later chapters of FO and CB.)

Orgone energy*: the life energy discovered by Reich in the late nineteen thirties. (See FO and CB for details of the history of this discovery.)

Orgonotic*: excited by, charged by, orgone energy. (See CB.)

Orgonotic pulsation*: the orgasm formula+ occurring generally throughout the organism in many guises in harmony with the organism's needs. Reich held that a state of health was the capacity for orgonotic pulsation throughout the organism without interference from muscular armouring. (See CB.) Orgonotic pulsation is also seen throughout living nature, for example in the pulsatory motion of a jellyfish or a worm.

Orgonomy*: the science and study of the orgone energy.+

Parasympathetic: the side of the autonomic nervous system+ associated with relaxation, pleasure and sexuality. For example, pleasure slows down the pulse and breathing rate, both parasympathetically stimulated. Antagonistic to the sympathetic+ side of the ANS.

pH value: a numerical description of a solution's acidity, which ranges from 1 to 14. 1 is very strongly acid, 14 is very strongly alkaline or basic. A neutral value is 7. The actual number is the inverse logarithm of the concentration of hydrogen ions in the solution. If your maths doesn't rise to this, get by, just remembering that low = acidic, high = alkaline/basic.

Phase contrast: a type of illumination used in biological microscopes which allows us to see details otherwise invisible. The light beam is divided by a phase plate or ring before passing through the object on the slide and re-united after this. The two beams emerge slightly out of phase so that there is some interference and an image where there would not be one with a direct bright-field beam of light. (See any microscopy textbook for more details.) An important aid in orgonomic microscopy.

Philosophical Transactions: the journal of the Royal Society, founded in 1666 and still published today under the same title. When the journal was first published, *philosophy* meant what we now call *science*.

Pollen: tiny grains of powder (accurate name – pollen sacs) attached to a male flower's anthers, which contain even smaller particles, the true pollen grains, the male gametes that fertilise the female flower's ovum. These correspond to spermatozoa in animals. Pollen sacs are just visible to the naked eye as a fine dust. The pollen grains are only visible under a microscope.

Protozoon (plural protozoa): a single-celled organism such as an amoeba or ciliate.

Pulsation: rhythmic expansion and contraction. (See also orgonotic pulsation above.)

Radiating bridge*: an orgonotic link between two bions,+ seen sometimes in bion cultures. The bridge shimmers like sunlight on water and is presumably a highly-charged orgonotic bond between the two bions. (See CB.)

Rods: a form of bacteria described simply by their shape, as are other types of bacteria – cocci, spirillae, and vibrios.

Spallanzani, Lazzaro, (1729-1799): Italian priest, natural historian and microscopist. Sceptical towards claims that spontaneous generation+ occurred, he claimed to demonstrate these were erroneous and that the living forms survived because experimenters took inadequate sterile precautions.

Spontaneous generation: the origin of new living forms from previously inanimate material. The expression is hardly heard nowadays, but was common currency in early biology and even in the early nineteenth century when many biologists still believed that it occurred. It is always assumed that Pasteur's famous experiments with the flasks and swan-necked tubes proved decisively once and for all that spontaneous generation does not occur in nature.

Sympathetic: the side of the autonomic nervous system+ associated with anxiety and contraction. For example, anxiety raises the pulse rate and blood pressure. Antagonistic to the parasympathetic+ side of the ANS.

Teleology: the assumption that a process in nature has a purpose or end goal. This was taken for granted prior to the growth of mechanistic science in the nineteenth century. Darwinism assumes that no natural process has any purpose and that nature, the origin of life, and evolution are purely random and accidental. Any teleological assumption, like vitalism,+ is anathema to modern science.

Tension-Charge formula*: the title of chapter 1 of *The Bion Experiments* and Reich's early description of the orgasm formula+ and the process that led to the origin of the bions.+ (See BEOL and FO chapter VII.)

Vesicle: a small blister of fluid surrounded by a membrane. Reich described a bion+ as an energy *vesicle*.

Vitalism: the belief that life processes involve some special organising principle active within living nature, making it different from non-living nature. It was a common belief before the rise of modern science in the nineteenth century. It is now the ultimate insult to throw at any scientific theory that suggests that there may be such an organising force. Reich's orgonomy,+ if it were known about by scientists, would be dismissed as *vitalistic* nonsense. Orgonomy gives vitalism a scientific foundation.

Work democracy*: a concept developed by Reich as a response to his negative experience of party politics and his realisation that political parties and programmes could not effect profound, life-affirming social change. Work democracy is not a political ideology or programme. It is simply what happens when a group of relatively healthy people engage in a common necessary task with no authoritarian interference from outside. Any authority generated during the task is spontaneous and originates in a rational need for it in order to attain the group's work goal and in the expertise of those exercising it at the time of a particular task. Authority in work democracy is, therefore, flexible and shifts with the needs of the job and there may be none at times. It presupposes a non-authoritarian character structure in those involved.

Index

A few living writers appear without dates. I could not find them after moderate efforts.

Abernethy, J (1764-1831), 46.
active molecules, 1, 2, 11, 12, 13, 17, 18, 19, 32, 37, 38, 47, 66, 69, 85, 111.
agglomerations of, 16, 61, 64.
Adams, George, (1750-1795), 62, 70.
Additional Remarks, (Robert Brown, 1829), 4, 12, 19, 40, 59.
agents provocateurs, 43.
air-germs, 28.
Al-Khalili, Professor Jim, (1962-), 69.
amateur scientist, 56.
Amici, G B, (1786-1863), 65.
amoeba, 26.
army, use of to control workers, 43, 44.
Atom, *The* (BBC), 68, 71.
attraction-repulsion in bions, 29.
Australia, 20, 21, 40, 43, 47, 113, 144.
autoclaving, 35.

Balfour, Eve (1899-1990), 58.
Banks, Sir Joseph, (1743-1820), 12, 40, 45, 46, 48.
Battle of Hastings, 42.
basalt, bions from, 19, 79.
Bastian H C, (1837-1915), 8, 14, 22, 35, 36, 39, 62.
BBC, 65, 68.
BBC-h2g2, 71.
Beagle, *HMS*, 110.
benches, 142-143.
Bennett, Philip, (1941-), 48, 50.
Benson, Ann, 9.
Bernal, J D (1901-1971), 35.
Benveniste, Jacques (1935-2004), 51.
Bioelectrical Investigation of Sexuality and Anxiety, (Reich, 1937), 10, 137.
bio-energetic charging, 103,
bion experiments, equipment for, 74.
Bion Experiments on the Origin of Life, *The*, (Reich, 1938), 10, 19, 22, 23, 32, 37, 38, 39, 62, 72, 93, 95, 111, 112.
bion experiments, 53, 68.
bion experiment, basic, 79,
doing, 72 et seq.
preparing first culture, 80-81.
bionous disintegration, 2, 13, 14, 26-27, 30, 91, 103, 110.
bionous motility, 5, 14, 27, 29-30, 64, 86.
bions, 2, 13, 14, 18, 27, 72, 84.
agglomerations of, 2.
appearance of, 24.
internal blue colour of, 24, 87, 99.
pulsation in, 87, 99.
bions, life-span of, 109.
Blake, William, (1757-1827), 37.
bone-meal, bions from, 79.
Bowley, Professor Roger (1946-), 65, 71.
breathing, 26, 68.
brick, bions from, 29.
Brief Account of Microscopical Observations (Robert Brown), 2, 114-126, and passim.
brightfield lighting, 84, 139.
British Museum, 5, 12.
Brown, Robert (1773-1858), 1, 3, 7, 11 et seq, 14, 15, 22, 23, 32, 41, 46, 48, 53, 57, 59, 64, 66, 67, 68, 70,

72, 87, 93, 96, 98, 103, 104, 105, 106, 110.
A Brief Account of...(1828), 1, 11, 12, 14, 17, 18, 19, 22, 23, 114.
text, 114-126.
Additional Remarks, 4, 12, 19, 40, 59, text, 111, 126-131.
apparent retraction, 4, 19, 20, 40 et seq, 48.
diary, 20-21, 23.
discovery of plant cell nucleus, 13.
distortion of his pollen findings, 59.
experimental controls, 16.
return from Australia, 1805, 43.
'slips', 16, 20.
persona, 21, 40, 47, 48.
pollen experiments, 47, 98-106.
Prodromus, 40.
Brownian motion, 2, 3, 5, 7, 11, 13, 16, 18, 20, 28, 29-30, 34, 37-38, 60, 62, 63, 65, 66, 67, 69, 80-81, 105, 107.
Brownian Motion in Clarkia Pollen (Ford), 64, 71.
Brunel Microscopes Ltd, 7, 72, 86.
Brunel SP40 microscope, 85,87, 99.
Brunel SP100 microscope, 76, 85. 87, 99, 108.
Brunel SP150 microscope, 85, 87, 88, 98.
Bruno, Giordano (1548-1600), 37.
Buffon, Count de (1707-1788), 4, 19, 41, 59, 62, 63, 66, 67, 69.
Bulloch, William (1868-1941), 22.
buying a microscope, 138-141.

Cancer Biopathy, The, 23, 24, 31, 32, 38, 70, 72, 93, 95, 97, 111,112.
carbon, bions from, 24, 28.
chalk, bions from, 80.
Chambers, Robert, (1802-1871), 24, 31.
character analysis, 26.
charcoal, bions from, 28.
Charmouth fossil shop, 106.
Clarkia pulchella, 64, 104, 105.
clay, bions from 28, 79.
Clegg J (1909-1998), 37, 39.
coal, bions from, 17.
coarse-focus stop lever, 83.
coccus, cocci, 27.
Combination Act (1825), 44.
common functioning principle (CFP), 47-48.
condenser, microscope, 77, 86.
'contamination' as explanation for bions, 90.
control experiments, 28.
Cook, James (1728-1779), 12.
copper filings, bions from, 110.
Cornwell J (1940 -), 71.
coverslip, microscope, 79, 82-83, 90.
Crimean War, 43.
criticism, right to, 6
cysts, 26.

darkfield lighting, 77.
Darwin, Charles, (1809-1881), 11, 12, 13, 20, 22, 23, 40, 49, 58, 63, 110, 112.
Darwinism, 56.
Davidson, Dean, 112.
De Duve, (1917-), 58.
democracy, work, 48, 50.
DeMeo, James, (1949-), 38, 58, 112.
Dobell, C (1886 - 1949), 22.
Döring, D, 18
Dorset, 106.
drawing microscope objects, 75, 91.

ebay, 5, 6, 72, 139.
Einstein, Albert (1879-1955), 3, 59, 60, 61, 69, 70.
Eldon, Lord Chancellor (1751-1838), 46.
empty magnification, 29.

English Civil Wars, 43.
Ether, God and Devil, 50.
evening primrose, 64, 104, 105.
Evolution of Physics, The (Einstein and Infeld), 59, 60.
executions, 47.
exine, 101.
expansion and contraction, 26.
experimentation, 57.
eyepieces, microscope, 76, 85.

falling wages, 44.
Farley, John (1936-), 38, 39, 45, 49.
FDA, tests of accumulator, 30.
feldspar, 17.
fertilisation, plant, 104.
field diaphragm/iris, 77, 83.
filings, metal, making your own, 80.
flaming slide, 83.
fire-precautions 107.
Fleck, Ludwik (1896-1961), 13, 22.
Flinders, Captain (1774-1814), 20.
focussing slide, 83.
focus stop-lever, 77.
focus wheels, 77.
forceps, coverslip, 79.
forceps, fine curved, (NHBS), 97.
Ford, Brian J (1939-), 7, 41, 49, 63, 64, 70, 71, 111.
  replication of Brown's pollen investigation, 64.
Fortean Times, 54.
fossilised wood. 16.
fossils, bions from, 79, 106.
fossils, grinding, 106-107.
franchise, 44.
France, war with, 44.
Freedom of Information Act (USA), 30.
freesias, pollen and bions from, 103, 104.
French Revolution, 45.
freshwater microscopy, 72.
Freud, Sigmund (1856-1939), 25.
Fuchs T, 58.
Fuller's earth, bions from, 79.
functional thinking, 135.

Gallileo (1564-1642), 11, 13, 63.
Gardner M (1914-2010); 32, 58.
geologysuperstore.com, 141.
Gillray, James, (1757-1815), 44.
glass, bions from, 17, 109-110.
glossary, 145-152.
gneiss, bions from, 29.
granite, bions from, 19, 29.
grass, bions from, 27.
grass-infusion experiment, 28, 88.
  extra items for, 88.
  observing, 90-96.
  attaching grass to slide, 90.
  bions appearing from, 94.
  autoclaved control experiment, 96.
grass preparation, 28.
Great Terror, The, 45.
Greenfield, Jerome (1923-2003), 32.
Gribbin, John (1946-), 58.
grosse Kosmos-Buch der Mikroskopie, das, (Kremer), 66.

Habeas corpus repeal, 44.
Haldane, J B S (1892-1964); 35.
Haran, Brady, 65.
Harris Library, Preston, 9, 10.
Harris, Philip, Ltd, 72.
Hartley W G (1913-?1999), 71.
Heisenberg, Werner (1901-1976), 69.
Henshaw, Carol, 9.
Higgins, Mary, (1925-), 8.
himalayan balsam, 104.
History Today Companion to British History, 49.
Hitler's Scientists, (Cornwell J), 69.
Hooper, Judith (1949-), 8, 10n.

hostility of science to life energy concepts, 51, 52, 53, 55, 56, 57.
Hudson L (1933-2005), 71.
Huxley, T H (1825-1895), 46, 62.

illuminator, 77.
Illustrated Dictionary of Science (Usborne), 66, 69, 71.
Industrial Revolution, 44.
industrial unrest, 44.
Infeld, Leopold, (1898-1968), 60, 61, 69, 70.
informers, 43.
Ingenhousz, J, (1730-1799), 63, 70.
Investigator, *HMS*, 20.
iris diaphragm (on microscope), 77.
iron filings, bions from, 19, 27, 28, 79, 97-98.
Isaacson, Walter, (1952-), 61, 69, 70, 71.
Isle of Mull, 29.

Jacobinism, 44.
jellyfish, 26.
Jones P, 32.
Jupiter Botanicus (Mabberley), 22, 49, 50, 63, 71.

Kettlewell, Bernard, (1907-1979), 8.
kinetic theory, 66.
Köhler illumination, 77, 140.
Kremer B P, 71.

laboratory, improvising, 142-143.
laboratory suppliers, 138-143.
Lacey J, 111.
Lancashire rising, 1826, 43.
Larousse Dictionary of Science and Technology, 111.
Lavoisier (1743-1794), 45.
Lawrence, William (1783-1867), 46, 50.
Leeuwenhoek, Anton van (1632-1723), 8, 14, 67.
Leibnitz, 67.
Leitz microscopes, 6, 87, 88, 99.
libido, 25.
libraries, public, 9, 24-25.
life energy/force, (see also orgone energy), 33, 56, 57.
light source, 77.
limestone, 18.
lumination, orgonotic, in vacuum tube, 52.
Lyte C, 22.

Mabberley, D J (1948-), 20, 22, 40, 43, 48, 49, 50, 63, 64, 71, 104, 112.
machine breaking (1812), 43.
Mad, Bad and Dangerous People, A? (Hilton), 49.
Maddox, John (1925-2009), 55, 58.
Madigan M T, 58.
magnification changer, 76, 85, 91.
magnification, high, 29, 76, 86, 87, 88, 91.
Marten, Dr Benjamin, 14.
materials, preparing, for bion experiments, 80.
materialism, 41, 47.
mechanism, 51, 53, 54, 58, 68.
Mendez, Armando, 9.
metal, filings, bions from, 79.
mica, 17.
micrometer, stage, 72.
focussing, 72-73.
microscope, compound, 21.
microscope parts, 78.
microscope, research grade, 72.
microscope, secondhand, 7, 139-140.
microscope, simple, 21, 63, 64.
microscope, single-lens, 63.
microscope specification, 72, 73.
microscope, stereoscopic, 98.
midwives, 68.

Millham, John., 141.
Milton, Richard (1943-), 23, 58.
monads, 67.
moss, bions from, 27.
motility, bionous, 5, 10, 14, 20, 29, 30, 46, 63, 64, 81, 82, 84, 88, 99, 105, 106, 108, 113, 114, 136.
motion, pulsatory, 20.
motion, vermiform, 20.
muscular armouring, 26, 51, 68.
mysticism, 51, 55, 57.

Natural History Museum, 16, 20.
Natural History of Pollination, The, (Proctor et al), 104, 112.
'natural order of things', 56, 57.
Natural Organization of Protozoa, The, (Reich), 48.
Natural Organization of Work, The (Reich), 48.
natural selection, 54.
Nature (journal), 51, 55.
Nazi Germany, 2.
Needham, John Turberville (1713-1781), 4, 19, 38, 41-2, 49, 59, 62, 66, 67, 70.
needles, 76.
Neill, A S (1883-1973), 28, 33, 35.
Neue Untersuchungen zu den Seesandbionen von Wilhelm Reich, (Palm and Döring), 112.
Newton Isaac (1642-1727), 11, 13, 63.
Nexus, 54.
Nomarski DIC, 7.
Norway, 37.
objective, microscope, 74, 77.
obstetricians, 68.
Of Moths and Men (Hooper), 8.
oil immersion, 85, 91, 98.
Oldfield R, 111.
Olympus microscope, 6, 76, 84, 87, 99, 108.
organge peel, bions from, 29.
organic molecules (De Buffon), 4, 59.
orgasm formula, 13.
orgone accumulator, 72.
temperature rise in, 53.
orgone energy,
denial of, 52, 57.
discovery of, 30, 52.
orgone energy in atmosphere, 53n,
orgone therapy, 52.
orgonomic functionalism, 135.
orgonomy, 33, 35, 68.
orgonomy, further study of, 143-144.
orgonomy in the world today, 143-144.
orgonotic pulsation, 13.
origin of life, 55.
Origin of Species, *The,* 12.
Oslo, 2, 34, 93.

Palm, Monika (1951-), 18,
paraffin wax, 89.
paediatrician, 68.
pastels, drawing, bions from, 80,
Pasteur, Louis (1822-1895), 55,
swann-neck flask experiment, 55.
Penguin Dictionary of Physics, 111.
Perrin, J B, (1870-1942), 3.
phase contrast lighting, 77, 84.
philosophes, 45.
pH value and bion growth, 18, 28, 29, 87.
pit-coal, 18.
Placzek B R, *Record of a Friendship,* 32, 38, 39.
plasmoids, 19.
Plato, 67.
'playing god', 57.
pollen experiments, 4, 98 et seq,
pollen grains, 15, 16, 64, 65, 66, 69,
bions from 101,
pollen, freesia, 103,

pollen, hive, 99,
pollen sacs, 66, 69, 100, 104-105.
Pope, Natalie, 9.
potato, bions from, 29.
potassium chloride, 98.
Pouchet, Félix (1800-1872), 39.
Priestley, Joseph (1733-1804), 45.
primary needs, 52.
protozoa, 36, 48.
pseudonyms, 33.
Psychoanalysis, 25,
pulsation, 26.
pulsatory motion (Brown), 16, 48, 64, 69.
Pulse of the Planet, 33, 38.
Purves W K (1934-), 58.
Puschkarew B M, 39.

quartz, 17.

radiating bridge, 30, 85.
reaction, political, in early nineteenth century England, 43.
red heat, 27, 81.
Reform Act, (1832), 44.
Reich Wilhelm (1897-1957), 1, 2, 3, 9, 14, 24 et seq, 25, 26, 28, 30, 34, 53, 57, 68, 93, 106, 110, 142.
  Beyond Psychology, 31, 38, 142.
  bibliography, 137-138.
  The Bioelectrical Investigation of Sexuality and Anxiety, 31.
  The Bion Experiments on the Origin of Life (1979), 10, 19, 22, 23, 32, 37, 38, 39, 62, 72, 93, 95, 111, 112.
  The Cancer Biopathy (1948), 23, 24, 31, 32, 38, 72, 93, 95, 97, 111, 112.
  Character Analysis, 32.
  Cosmic Superimposition (1953), 50.
  Ether, God and Devil (1953), 50.
  Die Funktion des Orgasmus (1927), 25, 31.
  The Function of the Orgasm, (1942), 24, 31, 32, 57.
  single iron-filing experiment, 96.
  text of extracts, 132-136.
Reill P H, 58.
Revolution, French, 45.
revolution, fear of in England, 43-44, 47.
rheostat, 83.
rice, ground, bions from, 79.
rising food-prices, 44.
rock-dust, bions from, 79.
rock-dust, suppliers, 79n.
Royal College of Surgeons, 46.
Royal Society, 41.
Rushton, Sharon, 46n.

safety precautions, 80.
sand, bions from, 28.
sand, making your own, 80.
Schleiden, Matthias (1804-1881), 21, 23.
Science, (Dorling Kindersley), 65, 71.
science, mechanistic, 15, 68.
Scottish Enlightenment, 12.
sea-sand, 18.
seed-germination experiment, 53.
SEER, (Sustainable Ecological Earth Regeneration), 79n, 142.
Semmelweiss, Ignaz (1818-1865), 37.
Sharaf, Myron (1926-1997), 31, 32, 39.
sharps, safe disposal of, 75.
Sheldrake, Rupert, 55.
Shelley and Vitality (Sharon Rushton), 46n.
silica, 17.
slide, microscope, 79.
slide, prepared, 72.

Slingsby, 143.
slate, bions from, 29.
Sloan, Philip, 63, 70.
Snyder, Maxwell, 8.
soot, bions from, 18.
Spallanzani, Lazzaro (1729-1799), 8, 39, 67.
Spanish Armada, 42.
spermatozoa, 62.
spinning motion, 85, 86, 102, 103, 104, 105, 108.
spontaneous generation, 36, 41, 45, 47, 55, 59.
spontaneous generation, conflation of, with democratic politics, 47.
spores, of ameobae, 26, 90.
stage, mechanical, 76.
stage, microscope, 76,
sterile precautions, 35,
Straus, Roger (1917-2004), 8.
streamings, bio-energetic, 26,
Strick, James (1956-), 8, 23, 39, 42, 70,
Sturman, Michelle, 9.
Subaseal, 28.
sundries, microscope, 74, 138, 143.
Sutton, Claire, 9.
syringes, 76.

teleology, 56.
tension-charge formula, 132.
test slide, 86.
test-tubes, 76.
Thoemmes Press, 1.
Trafalgar, Battle of, 43.
Transportation to Australia, 47.
Troy, Margaret, 9.
Tsiormpatsis, Stergios (1982-), 9.
Understanding and Using the Stereoscopic Microscope, (Woolnough), 112.

Van der Pas, Peter W, 62,
vacuum-tube, lumination of orgone-charged, 52.
vermicular motion (Brown), 61, 69.
vermiculite granules, bions from, 79.
vesicles, 2, 24, 27, 40.
Victoria, Queen, 43.
vitalism, 36, 42n, 54, 55, 57.
volvox, 95.

war with France (1803-1815), 44.
water immersion lens, 65.
water, memory of, 51.
wax mixture for grass-infusion slide, 89, 90.
well-slide, 79.
Woolnough L, 112.
'Workmate'® bench, 143.
World Wars, 43.
work democracy 47, 48.

Yorkshire, 43.
Yellow Pages, 79.
YouTube, 65, 68, 69.

Zeiss microscopes, 76, 99.